Skeeter Beaters

Skeeter Beaters

Memories of the South Pacific

1941-1945

Dennis Cline with Bob Michel

DeForest Press
Elk River, Minnesota USA

Published by:
DeForest Press
P.O. Box 154
Elk River, MN 55330 USA
www.DeForestPress.com
Toll-free: 877-441-9733

Richard DeForest Erickson, Publisher
Shane Groth, Editor in Chief
Cover by Michael Dear
Maps by Dennis Cline
Photo credits: Lois Cline p. 7; Bob Michel pp. 6, 137, 151, 218; Dave Minton pp. 181, 189, 207, 229, 232, 237, 245, 249, 255, 257, 267, 268.
Author photo, back cover: Carmen Cline

ISBN 0-930374-00-3
Printed in the United States of America

06 05 04 03 02 5 4 3 2 1

Library of Congress Cataloging-in-Publication Data

Cline, Dennis, 1949-
Skeeter beaters : memories of the South Pacific, 1941-1945 / Dennis Cline.
p. cm.
Includes experiences of Robert Michel, PhM3/c, USN Malaria Control Unit Cactus. Includes bibliographical references and index.
ISBN 1-930374-00-3
1. World War, 1939-1945—Health aspects—Oceania. 2. World War, 1939-1945—Medical care—Oceania. 3. Malaria—Oceania—Prevention. 4. Michel, Robert, 1923—Interviews. 5. United States. Navy—Medical personnel—Interviews. 6. United States. Navy—Medical care—History—20th century. 7. Mosquitoes—Oceania—Control. I. Title.
D807.O24 C55 2002
940.54'75—dc21

2002152171

—In Memory—
Sergeant William T. Cline, 1926-1999
Veteran, Pacific Theater of Operations
Company C, 1716th Aviation Signal Service Battalion, USAAC

While serving in the Philippine Islands in 1945, Sergeant Cline, like tens of thousands of other U.S. servicemen, fell prey to an infected sand fly contracting the relatively obscure tropical parasite Leishmaniasis. Cline suffered from boils and skin lesions common to the disease until 1958 when the parasite finally ran its course.

D.C.

Many thanks to Bob Michel
who shared his stories so freely.
D.C.

Bob Michel survived the war in the Pacific and shortly thereafter went to work as a Page for NBC in New York City. Over the years, Bob stayed with broadcasting, hosting numerous Radio & TV shows throughout the Northeast United States. Bob stayed for years with KDKA Radio in Pittsburgh under the stage name Bob Tracey. Appearing in over a dozen Hollywood films and theater productions spanning the past four decades, Bob, though officially retired today, is still active in the business. He recently appeared in a major motion picture released in January 2002.

This is in memory of my mom, Edith Rowland Michel,
whose strength and love were always there for me.
Robert Michel, August 2002

CONTENTS

FOREWORD

The freedom that each generation of Americans enjoys and generally takes for granted is founded in the deeds and desires of those that came before them. The generation that fought and died in World War II consisted of ordinary people who accepted, as their duty, the compromises and sacrifices required in securing freedom for those that survived. Our parents, grandparents, and relations are living connections to that history and legacy of freedom.

History to these veterans is a vivid set of highly personal experiences, not dull facts on the page of a book that nobody opens. History, through them, is given life and made all the more relevant to their families and descendants in a very personal and meaningful way. The author's conversations with two surviving members of the original Skeeter Beaters and a handful of other veterans of South Pacific campaigns took place almost sixty years after the events themselves. How fortunate it is for their families, friends, and fellow citizens that their story has been told and preserved while it is yet possible.

They, like countless others, performed without fanfare or notoriety and returned to nurture the free society in which we now flourish. It is appropriate that we, who have a share in their legacy, do honor to their courage and sacrifice and strive to emulate their dedication to a higher cause.

Rodney A. Jaroch

PREFACE

I first met Bob Michel (pronounced Michelle—just like the girl's name) in early August, 2000. At the time, I was thoroughly engrossed in the literature of World War II in the Pacific and even more so with the Guadalcanal campaign. Anyone who has been bitten by the subject is aware not only of its significance to U.S. history but also the broad range of material available on the subject.

I felt that I had surpassed, at the very least, all of the primary and most of the secondary source materials available and had moved on to the Military Branch Historical Records and Archive documents. I had also devoted many hundreds of hours seeking out veterans of the Guadalcanal campaign and pestering them to no end with questions of endless minutiae. In no time, I was convinced that my research had unearthed enough verifiable material to take a stab at adding to the record.

Bob Michel, like many other World War II veterans or their families, had posted a listing on the Internet—name, email address and a brief description of his World War II service. Bob described being a Naval Pharmacist Mate assigned to a Malaria Control Unit made up of an entomologist, two medical doctors, and thirteen enlisted personnel—all Pharmacist Mates like himself. Arriving shortly after the first Marine landings on Guadalcanal in August, 1942, Bob explained how his unit served throughout the course of the campaign until he and his buddies were individually transferred off the island during the months of January and February, 1943. Truthfully, a malaria control unit didn't sound that exciting to me when compared to the stories we grew up on of Marines and sailors locked in deadly combat with the enemy. My initial interest in contacting Bob, however, was to develop an additional source who might be able to provide a firsthand account of day-to-day life behind U.S. lines, inside the Lunga Perimeter. I thought Bob Michel could answer a few questions, so I sent off an introductory email.

Within a day, I received a reply from Bob, something to the effect, "I don't know how much I remember, but I would love to help you with your project." Excited by the quick response, I was not quite sure where to start with Bob but wanted to let events run their course. I quickly read everything I could get my hands on about malaria as it related to Guadalcanal, the South Pacific, and its impact on operations in the Pacific Theater in general. I put together what I thought was a pretty good list of questions and fired them back to Bob to see what he had to say. Quite frankly, I thought that would be the end of my correspondence with Bob.

Another day passed and I found a second, lengthy email from Bob in my Inbox. "Some of your questions are not relevant…naturally so, because you are not familiar with the function of Malaria Control. Let me explain." Reading further, I must say I had no idea who this guy was or what he was all about. I would ask Bob what I believed was an intelligent, historically significant question about the Guadalcanal campaign and Bob would send a reply so fast I thought he was just sitting around with nothing to do but wait for my emails. The only problem was, his answers to my questions were not quite what I was expecting. After brushing aside what I thought were thoughtful, "scholarly" questions, Bob would send two or three pages of the wildest stories I ever heard. Here I was reading Bob's neatly typed stuff and wasn't sure whether I should laugh or cry. His was a Guadalcanal that defied the historical record, the literature, and my sense of propriety all at the same time. I even started harboring doubts as to whether this guy's memory still served him.

Quick calculations led me to the conclusion that Bob had to be somewhere close to eighty or possibly, even older. Almost sixty years had passed since the campaign and I reminded myself more than once that time can play cruel tricks on a man's memory. This guy, though, regardless of his age, exhibited more than a pretty fair knowledge of computers, email, and the wide world of the Internet. All things considered, my curiosity was piqued and I was having one heck of a good time with Bob. As long as he wanted to keep telling the stories, I was more than happy to listen and see where they led.

This much I did know, however. By May of 1942, Japanese forces had eliminated any significant U.S. presence in the Pacific. British, Dutch, French, and German colonial possessions from India to Australia had fallen and were now part of the Japanese Greater Co-Prosperity Sphere. Virtually 1/7 of the surface of the entire world now lay in Japanese hands. In the South Pacific, vital communication and shipping lines between the United States, New Zealand, and Australia lay directly in the path of Japan's eastward advance through New Guinea and New Britain. With thousands of islands spanning the sea between the rich oil fields of the Dutch East Indies and French New Caledonia, the Japanese Navy steamed steadily forward, building island air bases along their path covering their rapidly expanding supply lines.

In May, 1942, Japanese troops landed on the obscure little island of Tulagi in the British Solomon Islands. Construction of a seaplane base along the shores of the island was soon completed. In just a matter of weeks, work commenced on a larger, land-based airfield on the northern plains of the island of Guadalcanal, 25 miles south of Tulagi and the largest island in the Solomons. Once completed, Japanese bombers would be capable of striking the New Hebrides and New Caledonia, isolating New Zealand and Australia in the process.

From December 1941 through June 1942, the U.S. Navy, undermanned, ill equipped, and poorly supplied, found itself fighting a holding action from the Aleutian Islands in the North Pacific to French New Caledonia in the South Pacific. Bouncing back and forth across more than 6,000 miles of open sea, just four aircraft carriers and a handful of cruisers and destroyers had already bombed the Japanese mainland, engaged in the first great carrier battle of the war in the Coral Sea, and had stopped Yamamoto's mighty fleet at Midway. Now, aerial reconnaissance over the

Solomons turned up evidence of a nearly completed airfield on Guadalcanal. With this knowledge, U.S. military planners decided the days of defensive operations were over and the time had come to take the fight to the Japanese. The airfield on Guadalcanal would prove to be the farthest point reached by the Japanese in their eastward expansion and the turning point in the war in the Pacific. Unknown to either of the combatants at the time, the stage was now set for the greatest combined land, sea, and air campaign of the entire Second World War.

I also had a pretty good understanding of the sequence of events that followed late in the summer and early fall of 1942.

On August 7, 1942, transported by the U.S. Navy from Wellington, New Zealand, elements of the 1st Marine Division landed on Tulagi and Guadalcanal. The Marines met fierce resistance on the island of Tulagi but remarkably, met no resistance when they initially stormed ashore on Guadalcanal. The Japanese, however, wasted little time in mounting retaliatory air strikes that afternoon against the supply and troop ships lying just offshore. Fearing the worst, Navy and Marine personnel worked around the clock unloading men and supplies. Just two days later Navy intelligence learned of an approaching Japanese naval force. Hauling anchor, U.S. troop and supply ships, followed by their escorts, hastily departed the landing area. With landing operations interrupted, the Marines ashore were left with what they had—literally. Every item necessary for their survival was not only in short supply but, in many cases, critically lacking. Capture and completion of the Japanese airfield now became the number one priority essential for their defense as well as the only reliable link in their already precarious supply lines. The siege of Guadalcanal had begun—a siege which tested the mettle of every man involved and would out last in duration even the siege of Stalingrad which was just heating up half a world away.

Now, along comes Bob telling me stories of Malaria Control Unit Cactus, the first malaria control and eradication unit ever attached to the 1st Marine Division and to operate in combat on Guadalcanal. Flown first class into Henderson Field just days after the initial assault on the island, the sixteen original members of MCU Cactus stepped off one of the earliest flights ever cleared into Henderson Field. As Bob told it, there were just sixteen guys in that first unit who were armed with a letter from Admiral Nimitz himself, answerable only to the highest levels of the First Marine Division brass.

Bob had me hooked to say the least because the only reference I was able to find regarding malaria control efforts during those first few months of fighting on Guadalcanal stated that malaria control personnel were not even allowed on the island until mid—November. How wrong that proved to be.

Bob's story and the story of malaria control in those early days of the war in the South Pacific is a great story. It started out as a lesson for me in history made more real by my personal interaction with Bob. It would have been easy to rewrite everything Bob told me in the style or format of another war memoir or biography. The fact is, however, that would only relegate the story to the nitpicking of scholars and professional historians. You see the story of the Skeeter Beaters is not a scholarly work—it's just a couple of guys sitting down and telling war stories. It's just Bob telling me what he remembers when he can, and what he has forgotten when he can't. The story is jumpy and sometimes scattered just as Bob told it to me.

Sometimes I had a hard time fitting what he had to say into my historical perspective. Many times I had to do some serious research just to keep up with him. I found that I was for the most part woefully illiterate in South Pacific geography and truly misinformed or at the very least uneducated in much of the nuance of the earliest U.S. initiatives in the South Pacific.

All in all, this is Bob's take, more outrageous and more remarkable than what any writer could dream up. The only additions I have made to Bob's story are historical and geographical sidebars, as well as occasional remarks and related memories shared by other veterans of the campaign. These additions have been included to provide perspective for the reader in better understanding places and larger events that were taking place within the scope of Bob's narrative.

To be sure, there are probably more than a couple of inaccuracies. This is not the story of how we won the war in the Pacific, but more the story of, "What did you do in the war, Daddy?" This is one man's memories of truly extraordinary times and how that very ordinary man remembers those times today almost sixty years later.

For me, Bob Michel's story was like sitting down to read **The Adventures of Tom Sawyer**, **Mutiny on the Bounty**, or **Revolt in the Desert** under a blanket at night with a flashlight after the lights had been turned out. Bob's story is a story of family, friendship, adversity, dedication, persistence, adventure, and hardship—oddly enough made possible by the wondrous technology of the Internet. It is also a story that would not be complete without the additions of many others who enthusiastically opened windows into their lives and contributed selflessly in sharing their memories of life changing times in our history.

When Bob ran into Dave Minton in the fall of 1941, the seeds of a friendship that would outlast 2 wars and 60 years were sewn. Serving together at the Brooklyn Naval Hospital for their initial medical training and later in MOB5 and Malaria Control Unit Cactus, Dave Minton and Bob Michel eventually wound their way across the South Pacific, serving together on New Caledonia, the New Hebrides, and Guadalcanal, separated only when the original MCU Cactus was disbanded in early 1943. Dave Minton went on to participate in two combat landings with the U.S. Army's 43rd Infantry Regiment on both Rendova and Munda rejoining Bob once more for a real knockdown liberty in New Zealand. Separated again until they were ordered back to the States in the spring of 1944, Dave and Bob hooked up on the same troopship, sailed together under the Golden Gate Bridge and headed cross country on one of the wildest train rides you can imagine. Dave Minton later rejoined his old outfit, the 1st Marine Division, in 1950 serving with distinction as a Company Corpsman as they battled their way out of the Chosin Reservoir. Sixty years—more than a lifetime for many—and yet Bob and Dave remain as close today as they were on their first liberty in New York City in 1941.

Before going too much further with Bob's story and that of malaria control, I think it now only proper to formally introduce the other major players who make up so much of the ensuing narrative.

Early on in my research, I was introduced to Tom Croft, a most remarkable man who joined the Navy in January 1941 to beat the draft but soon found a home at sea. Planning to make the Navy his career, Tom served aboard the USS Arizona in Hawaii before the events of December 1941. Transferred to the destroyer Monssen, Tom became a true "Rough Rider" with a ringside seat on one of the most unbelievable cruises in any Navy's history. Guarding convoys in the icy waters of the Northwest Atlantic, providing escort for the carriers during the super secret Xanadu mission led by Jimmy Doolittle to bomb Japan, racing down to the Coral Sea and back to the Battle of Midway, Tom loaded 5" shells in the Number 1 Turret on the bow of the Monssen as fast as his reach would allow. With no relief, Tom and the Monssen participated in the Doolittle Raid, the Battle of Midway, the invasion of Tulagi and Guadalcanal (August 7, 1942), survived the Battle of the Eastern Solomons later that month, and continuously provided covering fire and support to the beleaguered Marines ashore. In just less than 3 short months, Tom's dream of a career in the Navy ended forever on the burning deck of the USS Monssen when she was raked by 47 Japanese naval shells on Friday, November 13, in the Second Naval Battle of Guadalcanal. Though 85 years young today, every minute detail of time, place, and event have not escaped Tom.

William Ricciardi, a career Navy man serving aboard the USS Crescent City during her travels in and out of the Solomons carrying troops and supplies to Guadalcanal, added a wonderful perspective to the transient nature of life at sea and ashore. While very forthcoming with his time for this story, he asked that whatever he shared must be included in total. He was clear in his request that the whole story be told with no editing. "If you're going to write about what went on, don't pussy foot around the subject like everyone else. Tell it like it was or don't bother. And good luck, kid." During the course of the war, William was wounded four times participating in campaigns spanning South and Western Pacific battlefields and has lived for 54 years in a wheel chair as a result of his service to his country. Every story William told has been included with no holds barred. I am quite sure that if I were to call William today, he would have more stories to share.

One of the more fascinating and, to say the least, colorful characters I ran into during the course of writing this book I only refer to as "Marine Sergeant." When interviewed, he enthusiastically participated but asked that his name not be mentioned out of respect for those thousands who did not or could not participate in this undertaking. More than once over the course of a year or so, I exhausted the battery in my cell phone, so caught up in the stories of this man who served with the 11th Regiment of the 1st Marine Division. I just lost all concept of time. Here was a man who served throughout the war with the First Marine Division from Guadalcanal to Cape Gloucester to Guam, and served again in China for the better part of 1946 before returning to the States. Using the GI Bill to complete the education he missed during the war years, Marine Sergeant went on to become a successful airline pilot, flying 747s for TWA. It was Marine Sergeant who most succinctly said, "You know, the early part of my life was pretty rough, but things sure turned out well for me after the war."

Stewart Winterton, a Marine rifleman, survived four months in one of the world's most dangerous environments. He was pitted daily in hand-to-hand combat with the Japanese and was instrumental in describing much of the horror of life on Guadalcanal. Stewart's distinguished service with Chesty Puller's outfit, the 1/7, ended ignominiously when he was carried off the island on a stretcher with a life threatening case of malaria.

Bill Dickman, another pilot, flew transport planes for the Marine Corps originally as an enlisted man and later as an officer all over the Pacific and throughout postwar China, logging thousands of hours in military aircraft. Following a distinguished career, Bill retired a highly decorated Colonel. It was Bill who provided valuable information on the early efforts of Navy malaria control in the South Pacific and the long reach of their authority.

John Francis Richter, a Pharmacist Mate who filled one of the toughest jobs offered by the United States Navy—that of Company Corpsman—provided medical aid and moral support to the men of the 1st Marine Division throughout the Guadalcanal campaign. A member of the Navy but assigned to the Marines, John not only served the Marines, but lived their life as well. Guadalcanal, Cape Gloucester, and beyond, John later served in the Korean conflict and retired from the Navy having served so many, so faithfully. John's memories of the incongruity of life on the ground on Guadalcanal, as well as his knowledge of the ever-present dangers of disease, were an added bonus to Bob's.

Once I started interviewing veterans with an earnest, I was quickly overwhelmed by a very pleasant chain reaction—it seemed every guy I talked to had a buddy he thought could add more to the story. Such circumstances led to my introduction to Thomas Gross, and what a find he turned out to be. Thomas Gross was drafted into the service of the U.S. Army in January of 1941 for the compulsory term of one year. Most people today aren't even aware there was a military draft at that time, yet Gross was one of its victims. With little more than a month to serve when Japanese bombs fell on Pearl Harbor, Gross's tour of duty was extended for the duration. Gross ultimately served with distinction for over five years as both an enlisted man in the South Pacific and later as an officer in Europe. It was Thomas Gross's diary which identified the date and place of the horrific bombing in Purvis Bay in 1943 that Bob had such a strong memory of but could not accurately place in time. It was Gross who provided perhaps the most detailed view of an American fighting man's feelings toward the Japanese.

Lewis May, a graduate of the University of Washington, served with the United States Naval Reserve on active duty throughout the war and later became a prominent judge continuing to serve the citizens of this country until his recent retirement. It was Lewis who provided interesting anecdotal material to the participation of the Fijian Battalion during the Guadalcanal campaign.

Bill McLaughlin began his military service in 1936—a 16-year-old teenager with the now long forgotten CMTC, Civilian Military Training Camps program. Early on, Bill served with one of the U.S. Army's last active horse mounted cavalry units and later with the Americal Division in New Caledonia, Guadalcanal, and in the jungles of Bougainville. The final capitulation of the Japanese in 1945 found Bill a battle-hardened platoon sergeant serving with an elite Scout/Recon team in the Philippine Islands. I spent many pleasant hours reminiscing with Bill over memories passed on to me by my father of his days in the Philippines late in the summer and fall of 1945.

Unfortunately, Bill McLaughlin will be unable to see the work in print he so dearly encouraged. Bill McLaughlin passed away the summer of 2002 and will be sorely missed.

When I first spoke with William Schultz by phone, it took some convincing on my part to assure him I hadn't called to talk about his brother who served with the Navy and was captured by the Japanese after the fall of Corregidor. I explained that anyone who volunteered for special duty their first day in the New Hebrides and was handed a backpack sprayer for their enthusiasm was a man I had to talk to.

William volunteered as a construction specialist for the Navy SeaBees. Upon his arrival at Espiritu Santo in the New Hebrides, William was faced with much the same choice whether or not to volunteer for "special duty" as were Bob Michel and Dave Minton just two years earlier. He, too, soon found himself strapping on a backpack sprayer and hacking his way through the jungle for the next 19 months in search of the *Anopheles* mosquito. A young man, like many other hundreds of thousands, who was slated for the invasion of Japan, William served out his enlistment with the SeaBees to the war's end, destroying every item of heavy equipment in the U.S. military arsenal deemed too expensive to return to the States.

As referrals kept coming in and my network of contacts continued to grow, Robert Otto, an old destroyer sailor, was added to my daily contact list. Just as Tom Croft had contributed so much to the story of life in the Solomons from a sailor's view, Robert Otto proved no different. Serving aboard the USS Grayson in an open 20mm gun mount during the Battle of the Eastern Solomons, an awestruck Robert Otto, still in his teens, watched as a Japanese Zero roared so low over his gun that he was able to clearly see the pilot thumbing his nose at him. As a recruit with only two weeks of boot camp before being sent to the fleet, Robert went on to earn nine of the thirteen total battle stars awarded to the Grayson. Three years later he was wounded shortly after landing on Iwo Jima, ending forever his dream of a career with the Navy.

Following six long months of networking, interviews, and trips to the library, I was introduced to Arthur G. King, MD, whose military experience dated back to ROTC classes at Harvard College in the late '20s. Commissioned a 2nd Lieutenant in the Army Medical Corps upon his graduation from Harvard Medical School in 1930, Dr. King went on to serve as Task Force 6814 Medical

Inspector, later serving as Island Medical Inspector at Noumea, New Caledonia. It was Dr. King who, at the age of 94, proved to be the missing link in the elusive story of malaria control. He provided valuable insight into the startling ramifications of the sinking of the S.S. Coolidge while sharing so many wonderful stories of his 2-1/2 years service in the South Pacific.

Finally, crossing the Atlantic via cyberspace, I met Brian Taylor PhD, an Entomologist who, for years in the 1960s, was employed by the British Government in the capacity of Malaria Eradication Officer for the Solomon Islands. Dr. Taylor added perspective not only to malaria eradication efforts in the Solomons years after the war, but also insight into the local geography, ethnology, and scientific terminology I continued to struggle with. Though 5,000 miles distant, it was Dr. Taylor who finally identified and located the original Navy entomologists of Malaria Control Unit Cactus.

When Bob Michel, Dave Minton, and the rest of Malaria Control Unit Cactus stepped off an Army transport in August of 1942 onto the crushed coral runway of Henderson Field, they had absolutely no idea where they were or what was expected of them. Eighteen years young and incredibly naïve, while they didn't know it, they had just landed on a South Pacific island in the middle of one of the world's most hostile environments. Deployed to an island occupied by thousands of battle-hardened Japanese troops, Malaria Control Unit Cactus was about to enter into battle with a disease that had been endemic to the area for hundreds of years.

Given no training, no special equipment, and left to their own devices to scrounge for food, clothes, and shelter, they, like everyone else on the island, were expected to do their part in defeating the Japanese. Ill equipped and overlooked, the men of Malaria Control Unit Cactus went about their business enduring the same bombs, shells, and bullets as every other Marine and sailor who were more aptly trained for combat than they were. While other units were evacuated for rest and refitting, MCU

Cactus stayed on, serving longer on Guadalcanal than any other U.S. military unit at the time.

Today, Bob Michel refers dryly to that hell of almost 60 years ago with good-natured understatement, "Ours was by no means a glamorous job. It was hot, dirty work." Dave Minton down plays their role as well, "Bob and I had a job to do and we did it—simple as that. It was just a job like any other."

Job or not, hell or not, this is how Bob, Dave, and the others remember it after all those years. My only hope is that you enjoy the journey that Bob, Dave, and all the guys are about to take you on. If their story strikes a chord of curiosity in you to learn more about our mutual heritage, well then, I would say the guys did a good job.

ACKNOWLEDGMENTS

While compiling the stories, anecdotes, and perspective that make up the **Skeeter Beaters**, I quickly learned that telling the story was clearly not an enterprise of one. Early on I realized I would need the assistance of many. Now that the story is as complete as my research has taken it, many thanks are due those responsible for the right things that are included. Responsibility for any errors or omissions are entirely my responsibility and any criticism or blame for misquotes should be properly directed.

Every effort to contact members of the original Skeeter Beaters, MCU Cactus, Roses, Buttons, and MOB5 have been exhaustively played out over the past 2-1/2 years to no avail. Bob, Dave, and myself would love to hear from any parties with knowledge to the facts or who may have been primary participants to this tale and can be contacted through the publisher.

The patience, encouragement, and enthusiasm for the story of the Skeeter Beaters over the long haul of its development provided by my wife, Carmen, was truly the foundation upon which this book is built. My two sons, Sean and Joel, provided the integral impetus to keep going with their continued reassurances.

Pvt. John "Honest John" Minard, 98th Regimental Combat Team, father-in-law, future fishing partner, and all around problem solver, continually provided behind the scenes support that kept the project alive when the going got tough.

No work could ever go to print without the assistance of that one person willing to perform the painful, but necessary, "commaectomy" and the scribbled "Huh?" in the margins. For the **Skeeter Beaters**, Lisa Kurdziel DVM pulled no punches while spending hours explaining the difference between a vector and a reservoir.

I must include a formal thank you to Marj Michel who always seemed to be the one running for the phone at the most inopportune times in response to my incessant calls to Bob. Marj always unquestioningly shared her enthusiasm as the story unfolded.

There is one person, however, without whom the story of the Skeeter Beaters would never have been told and to whom I owe a special debt of gratitude—Dick Erickson of DeForest Press. It was Dick who truly believed in the project from day one. For well over a year, Dick provided the encouragement and assurance that Bob's story was indeed a good one and one worth telling.

Finally, I would like to personally thank all those named or anonymous who unselfishly provided their time, their stories, their insight, and an open door into their lives, without whom the elusive story of the Skeeter Beaters would remain yet untold.

INTRODUCTION

When the United States entered the war on December 7, 1941, every branch of its military was woefully unprepared for offensive operations anywhere in the world. The Navy and Marines in particular, were being systematically swept from the seas and off island bases all over the Pacific by overwhelming Japanese forces. It would take four years and the combined efforts of the military and American industrial might to prevail in a two-front war. It wasn't until the fall of 1943 that the minimum adequate supplies of food, clothing, guns, artillery, planes, ships, and men started arriving at the front lines in sufficient quantity to even make a difference.

What troops and equipment the military did send to meet the enemy prior to that time was, in many cases, simply what was available—outdated leftovers from the First World War. It was a precarious situation in which everyone had to make do and learn as they went, not only to win a battle here and there, but just to survive against an enemy better trained and equipped. Time was critical to Allied survival and it was the superhuman efforts of those first few men thrown into the front lines that ultimately bought the valuable time needed for the full weight of American industrial superiority to kick into high gear.

The United States Navy, facing the daunting task of rebuilding their shattered Pacific Fleet while at the same time, maintaining vital communications and supply lines with Australia and New Zealand, already knew that the health of their sailors and Marines would be in jeopardy as they moved into the South Pacific. As early as 1940, Navy Medical Corps planners launched an aggressive program designed to protect the health of its servicemen. They recruited and trained new Pharmacist Mates, formed Mobile Field Hospitals to care for expected casualties resulting from coming offensives, and organized malaria control and eradication units to minimize the risk of exposure to deadly jungle diseases. Just as every unit in the Navy was understaffed and overtaxed, so too were the men of the fledgling malaria control units first sent to the South Pacific. They too were forced to hold a line and buy time while critical shortages in manpower, training, and equipment were satisfied.

The first men who volunteered to "head up into the islands where the action was," men like Bob Michel and Dave Minton, were no different than the Marines they accompanied. The early members of malaria control units were thrown into battle improperly trained and poorly equipped, lacking the tools or technology that would bring them success. It is safe to say, that like many other military initiatives mounted in 1942, Malaria Control Unit Cactus was initially not that successful on Guadalcanal, nor were other malaria controls units very successful in their efforts early on in the New Hebrides.

However, by the fall of 1943, what the members of the first malaria control units had learned and had the time and resources to put into play would be put to good use throughout the remainder of the war in the Pacific. Once malaria control was relegated the importance it was due and units started to receive needed reinforcements, the equipment they needed—protective clothing, mosquito netting, advanced insecticides, and adequate medical supplies—and given the opportunity to disseminate what they had learned to front line troops and at every airfield, success in the war against disease was transformed from a dream to reality.

All in all, almost 9,000 men were incapacitated by malaria on Guadalcanal alone from August to December of 1942, and the casualty figures kept mounting well into the spring of 1943 after the Army had taken over operations. Just one year later, in the months from June to November, 1943, a mere 194 men were stricken with malaria during the Bougainville campaign.

Casualties due to malarial infection continued to decrease so drastically in subsequent Pacific campaigns right up to the end of the war that its effects were no longer deemed a factor in achieving military victory. Navy and Army malaria control personnel and dedicated units performed their tasks as well as any combat unit regardless of branch of service. It may never be known for sure just how many servicemen were able to stay active in combat, service, and supply due to their efforts.

Chapter 1

The Most Ferocious Enemy in the South Pacific--the Mosquito

"A mosquito landed at Henderson Field this morning and before we discovered it wasn't one of our planes, we had already fueled her up!" Unknown Marine, Guadalcanal, 1942

VECTOR BORNE DISEASES, of which there are dozens, are native to over 100 countries worldwide. A vector borne disease is a disease carried by insects that feed on blood. The most common of these vectors, or disease carrying insects, is the mosquito—most specifically the *Anopheles* and the *Aedes aegypti* species. The *Anopheles* is most noted for the spread of malaria; the *Aedes aegypti* for the spread of dengue fever.

Throughout the Pacific, over 75% of islanders live in rural areas and to this day are particularly vulnerable to the vector borne diseases. As isolated as we are in the United States from such perils, 45% of the natives of the Solomon Islands and the New Hebrides, today known as Vanuatu, are at risk. Unbelievably, the leading cause of death in children within this region remains vector borne disease.

In early 1942, there were three primary vector borne diseases that plagued the South Pacific and were of particular concern to military planners and strategists when the decision was made to stop the Japanese advance south and east across the Pacific. It was these three diseases, malaria, dengue fever, and filariasis that became the target of Navy and Army specialists as U.S. troops advanced into the area.

Perhaps the most widely known of the three diseases was and still is, malaria. Malaria is transmitted when the human body is bitten by a mosquito carrying the malarial parasite. The parasite is always transmitted to humans by the *Anopheles* species of mosquito, which most generally feeds at night.

The parasite is deposited by the *Anopheles* into the human blood stream and then makes its way to the liver where it takes up residence and multiplies into approximately 40,000 new parasites within just five days. The parasites travel back through the bloodstream releasing toxins along the way causing high fever, chills, and sweats. Once bitten, it only takes about two weeks for the first attack to take place. Since the parasites reside in the liver, it is not uncommon for the disease to recur months and even years after a person is first infected.

Unlike malaria which is caused by a parasite, dengue fever and dengue hemorrhagic fever are caused by a virus carried by the *Aedes aegypti* species of mosquito. The *Aedes Aegypti* mosquito feeds during the day, hence a day-biter. Dengue fever is also commonly known by natives and visitors to the area as "break bone fever" because of the violent tremors caused by fever. Severe cases diagnosed as hemorrhagic fever, often leads to internal bleeding and bleeding of the gums and nose. Left untreated, advanced cases of dengue fever can result in death.

Filariasis, the third most prevalent vector borne disease common in the South Pacific, is a worm infection. Once bitten by an infected mosquito, a larvae is passed from the mosquito into the victim's bloodstream traveling to the lymph system where it begins to develop into a worm. It takes approximately 6-12 months for the larvae to develop into worms. Once fully developed, the female worm then transmits more larvae into the bloodstream, starting the process of development all over again.

The most common example of filariasis infection in the United States is known to the dog lovers among us—heart worm.

The mosquito has been credited with laying whole armies low since the time of Julius Caesar's Roman civil wars. The impending conflict in the South Pacific would prove no different. Left unchecked, the tiny mosquito would prove a more deadly adversary than the full might of the Japanese Army.

Leading the attack, the female *Anopheles farauti* mosquito, indigenous to hundreds of islands in the Solomons and New Hebrides (Vanuatu), would soon play a major role in the outcome of hostilities on Guadalcanal. Taking no sides, malarial parasites rapidly infected both U.S. and Japanese troops in staggering numbers. U.S. troops occupying the islands were especially hard hit. Nine hundred cases of malarial infection were admitted to the hospital during the month of August, escalating to over 1,700 in September, 2,600 in October, 2,400 in November and more than 900 in just the first ten days of December. Total U.S. casualties during this period approximated 8,580 men—a number representative of 40% of all U.S. forces stationed on Guadalcanal during the first six months of occupation. If one were to look at the entire Pacific Theater of operations throughout the duration of the war, malaria was responsible for nearly five times the number of U.S. casualties than those caused by Japanese forces.

While the seriousness of the spread of malaria among U.S. troops on Guadalcanal is not to be diminished, Japanese troops were infected, probably to the man. Though U.S. military investment in ongoing malaria control efforts was puny at best, many of the troops benefited from mosquito nets, quinine, and the quinine synthetic, Atabrine. More importantly, the luxury of setting up camp on high ground and in open spaces close to the beaches, away from the mosquito's breeding grounds, saved many. The Japanese, on the other hand, constantly pressured by the Cactus Air Force and Marine patrols, were forced to live and fight in low lying jungle areas where the female *Anopheles* bred with impunity. The average Japanese soldier suffered unmercifully as a result of inadequate or nonexistent field hospitals, lack of medical supplies, absolutely no malaria control initiatives, and very few mosquito nets. One Japanese source states that as many as 2/3 of all Japanese fatalities suffered during the six month Guadalcanal campaign were a direct result of disease; malaria, beriberi, dengue fever, and dysentery. The majority of these fatalities was due to malarial infection.

To combat the spread of malaria on Guadalcanal, the 1st Marine Division included a squad sized unit designated Malaria Control Unit Cactus (MCU Cactus). The unit was hastily formed by Lieutenant Commander James Sapero at Mobile Naval Hospital #5 (MOB5) in Noumea, New Caledonia. The original MCU Cactus included Ensign Kenneth Knight—a Navy Entomologist, Dr. Robert Mount and Dr. Freedman, both MDs, and a handful of enlisted Navy Pharmacist Mates. MCU Cactus was subsequently dispatched to Guadalcanal on one of the first supply flights into Henderson Field, arriving within days of August 12th, 1942.

The significance of the Navy's Malaria Control Units seems almost implicit. When analyzing the casualty figures sustained by U.S. forces on Guadalcanal as well as those suffered by all military forces throughout the entire Pacific Theater of Operations, one is quick to understand how woefully understaffed and outgunned Medical Units were. Though significant in afterthought by today's standards, military leaders at the time lacked an understanding of the full impact of tropical disease and the resources they would need to battle them.

Even today, there are somewhere between 300 and 500 million cases of malaria of various strains reported worldwide annually. The prevalence of malaria in the Solomon Islands has not changed all that much since combatants vacated the area almost fifty years ago, and is even on the rise in some areas.

Though at peace once again with only Japanese tourists and businessmen in the islands today and the victorious Allies gone to their own pursuits, the indigenous enemy of native Solomon Islanders, the female *Anopheles* mosquito, still breeds in the area at will. With all the technological and scientific advances the world has seen since the end of the war, malaria eradication efforts will probably never be 100% successful in erasing the breeding grounds of disease carrying insects. An abundance of rivers, creeks, and lagoons, fed by torrential rains year round, continue to guarantee an environment in which the female *Anopheles* can breed and multiply. If, through some Herculean effort, every river, creek, and lagoon were drained, more puddles would continue to pop up following the rains like a bad case of measles, filling ruts on remote jungle roads whenever a truck or school bus passed by.

Malaria Control Unit Cactus, though well intentioned when it was formed, was doomed from the start to fight a holding action. Military commands, though well aware of the deadly threat to their troops from the mosquito, had no choice but to throw every asset they could muster against the Japanese. It will never be known for sure if any of the troops under siege at Guadalcanal benefited from MCU Cactus's daily patrols through the swamps and lagoons of the island. What is known, however, is that the techniques and equipment MCU Cactus developed in those early months delivered in spades in future Allied assaults throughout the South and Central Pacific. Trained in the use of bug sprays, the effectiveness of mosquito netting, and armed with medicines essential to prevention and treatment of malarial infection, casualties were eventually reduced by almost 98%.

It is surprising, therefore, that almost fifty years later the Malaria Control Unit Cactus has been all but forgotten except to its few surviving members. Lost in the shadow of much larger events being played out at the time, details of malaria control efforts are rarely mentioned in the vast literature on the South Pacific campaign.

It would eventually take the combined strength of 27,000 U.S. troops to push the Japanese from Guadalcanal and surrounding islands, saving Australia and New Zealand in the process. Until January 1943, however, a mere handful of men were thrown into battle on Guadalcanal against an enemy that accounted for more U.S. casualties than did all Japanese war efforts combined; this, in the campaign that halted Japan's expansion across the Pacific. A handful of men who were ordered to face an enemy that, to this day, still kills as many as three million people worldwide every year.

Robert Michel, PhM3/c, USN Malaria Control Unit Cactus. In order to get malaria, you must be bitten by the female *Anopheles* mosquito which has already bitten someone with malaria. The female *Anopheles* is the only one that transmits the parasite. Dengue Fever, often called breakbone fever, is transmitted by the Aedes Aegypti mosquito. These diseases are transmitted from an infected person to the mosquito and on to the next person. So we could have been bitten hundreds of times and not gotten the disease if the mosquito had not bitten an infected person first.

I think I first got malaria about two months after arriving on the Canal, but I can't be sure. I always took the Atabrine just as it was prescribed as a prophylactic along with the quinine if I came down with it. Yes, my eyes turned yellow, just like my skin.

One attack had my temperature up over 104 degrees and I was shaking so bad I couldn't hold anything. When I was finally admitted to the hospital there was an air raid going on my first night there. Marines had set up a 90mm antiaircraft gun emplacement right behind my bunk that was firing at the Japanese planes. I first realized it was there when they fired their first round at the incoming planes. That scared me. I don't think that it was more than twenty-five to thirty feet away from where I was lying in bed. I thought the damn thing was pointing right at me. I was so sick I didn't even try to get to a foxhole. I was sweating so badly from the malaria that the mattress I was lying on was soaked all the way through. The noise of that gun firing really scared me.

I ended up being kept in the hospital for two or three days, I'm not really sure. I guess, during much of that time, my fever was so high, I must have been in a partial state of delirium.

I don't really know when I finally got rid of the malaria either. I did have a recurrence later while I was stationed on the Florida Islands, but not too bad. I always had a supply of Atabrine and quinine with me to self medicate. I know I still had the bug when I finally got back to the East Coast in 1944. I think I had an attack as late as 1946 but know that I haven't had one since. Even after I got out of the Navy, I kept some quinine with me just in case I would have a recurrence. I must have had recurrences of malaria eight or ten times before it finally wore itself out.

As to why some got it and others did not, I can only give my opinion. At night when you went to bed, you sometimes might also have netting that was suspended over the bed by a wooden frame; that is, if you even had a bed. For the Marine sleeping in a foxhole with no netting, when he was bitten by a female *Anopheles* mosquito, he got malaria. His buddy lying right next to him might not get bitten. Plus, some of the troops there kept their shirts on and their sleeves rolled down and buttoned which reduced the chances of being bitten. It was a coin toss. Dave and I were in the islands for two years and I don't think he ever came down with it.

The only treatment for malaria was quinine and often many of us would give out the quinine instead of the substitute Atabrine as a prophylactic because you didn't turn yellow on the quinine.

Brian Taylor, PhD, Nottingham, UK. Onetime Government Entomologist, Malaria Eradication Programme, Solomon Islands. Malaria Eradication campaigns, as defined by WHO—the World Health Organization—never aimed to eradicate the mosquito vectors but to interrupt transmission, thus preventing new cases. Efforts to detect human disease carriers and use drugs to eliminate the parasite from them would, it was hoped, break the cycle and, so, eradicate the disease.

On the malaria parasite itself, in the Solomons there were, and still are, three species—*Plasmodium falciparum* (the killer), *Plasmodium vivax* and *Plasmodium malariae* (the form which recurs often years afterwards as relatively mild fever). *P. falciparum*, as noted the only killer, if survived, soon disappears from the sufferer. *P. vivax*, can reappear from a resting stage in the human liver for some 3-4 years after the initial infection. In terms of military impact, only *falciparum* would have a major effect, though the ill-effects of the other two would have been bad for an already stressed and fatigued soldier.

In the Solomons and Vanuatu (the New Hebrides), as in Papua New Guinea, the main vector species was, and still is, *Anopheles farauti*. Not in Vanuatu, two other species, *Anopheles punctulatus* and *Anopheles koliensis* also were malaria vectors. Indeed, *punctulatus* may have been more important on Guadalcanal as it breeds mainly in transient pools—such as the wheel ruts which must have been everywhere during the military activity.

The second, *koliensis*, is a curiosity. Prior to the U.S. forces' activity it was unknown—indeed, the name comes from its discovery at Koli, a village in the U.S. occupation area on Guadalcanal. The reference to dengue is interesting as it was not common, if known, in the Solomons in the postwar era, and the limited evidence suggested that *Aedes aegypti* was never common in the Solomons (especially on Guadalcanal where there was no real port prior to the U.S. Forces establishment of Honiara). Dengue is a disease better known from the (non-malarious) islands eastwards in the Pacific and in Southeast Asia. On the question of who gets malaria, there is certainly a variation in human "attractiveness" for mosquitoes due to a variation in sweat odor and perhaps surface body temperature. For a description of life on the Solomons in 1937-1942 the reader might like to find a copy of Dick Horton's book, **The Happy Isles**.

Chapter 2

The Beginning

August 1923-June 1942

Robert Michel, PhM3/c, USN Malaria Control Unit Cactus. My mother, Edith Rowland, was born on December 5, 1898, in the town of Neuderm, now called North Bergen, New Jersey.

Her father, Edward Rowland, was English. Her mother's maiden name was Anna Maria Teresa Wittnebert. She was German. Her grandfather and grandmother were native of the Tyrol district of Austria. They came to the United States from Europe before she was born.

Her grandfather or great grandfather, Peter Paul Diem, had been a Burgermeister in the town of Innsbruck, Austria. My mother's father (my grandfather), Edward, as I remember her telling me, was always pretty good at getting things easily, but always legally. He ended up with a good bit of property in and around Neuderm, where much of his wealth came from.

My mother, whose nickname was "Pete," also had four sisters: Florence, Hattie, Berdie, and Mildred. Two were older and one younger.

Aunt Florence, the oldest, was married to Ferman Haines and lived to be a hundred and one or a hundred and two years old. She lived most of her life on South Clarkson Street in Denver, Colorado. She had one daughter, Helen, who is still alive and living in the same house.

My Aunt Hattie, who married Harold Schuyler, and my Aunt Berdie, who married Fred Garlicks, both lived in the town of Rutherford, New Jersey. Sometime in the early '30s my Aunt Berdie moved to Hackensack. My Aunt Hattie's family moved to Ridgewood, New Jersey, sometime in the '40s.

My mother, Pete, married Lester P. Michel, who was my father.

I was born on August 11, 1923 in my mother's bedroom at 207 East Pierpont Ave. in Rutherford, New Jersey. I was delivered by Doctor William Williams, a general practitioner, who practiced medicine in that town.

My mother was her father's favorite. When she married, she had money and set my father up in the roofing business. I always assumed it was successful because he spent money freely. I remember, when I was about four years old, going outside one day to have a look at our new family car. It was a Willys-Knight with a rumble seat. My father put me in the rumble seat and drove us all around town.

I also remember one Christmas when I came downstairs and looked under the Christmas tree. There was this Lionel electric train underneath the tree on what they called standard tracks. What a beauty it was. We also had a dog whose name was Chong; a red Chow who was my guardian from the time I was in a playpen. Chong would lie next to the playpen out in the front yard and heaven help anyone who got too close. As she got older, she got sick and I remember my father using a .25 caliber lever action rifle to shoot her down in our cellar, putting her out of her misery.

Chong was replaced a short while later by a German shepherd named Rex. We had him for a few years until one day he just disappeared.

I think it was just after the market crash of '29 that we lost the house in Rutherford, New Jersey, and bought one on Vreeland Avenue in Bergenfield, New Jersey. It was just a little house that I think my parents bought for $3,400.00. It was brand new and one of the things I can still remember is my father in the kitchen frying potatoes and onions. The aroma is still with me today.

I remember going to a school in Bergenfield that was so old it was falling apart. I also remember my mother and I walking about two miles to a dental clinic to have my teeth taken care of. All my life, while she was still alive, she insisted that I take care of my teeth.

About this time, my father left us for another woman. The remaining family, my mother, my brother Doug, and I had no income—nothing. Of course, we lost that house in Bergenfield in short order. My Mother and my brother went to live with my Aunt Berdie in Hackensack and I went to live with my other aunt, Hattie, in Rutherford. Aunt Hattie's family, the Schuylers, had three children of their own when they took me in. Shirley was the oldest, Whitford was my age, and Harold Jr. was the youngest.

While I was living with the Schuylers, my mom got a job cleaning house for a woman on 2nd Street in Hackensack. Once in a while, I would get on my bike and pedal my way from Rutherford to 2nd Street in Hackensack, which was about six or seven miles. I would ride back and forth in front of the house where my Mom worked (I wasn't allowed to ring the bell because the woman didn't want my mother to have any visitors) hoping she would see me and come out for a few minutes. She could only come out for a few minutes though, because the woman she worked for was very strict about my mom working. She wanted my mom to work—not visit. The visits were, therefore, short and lasted only five or ten minutes, then I would go back to Rutherford. There was, at the time, home delivery of bread and milk, so I would sometimes hitchhike a ride up to Hackensack and then ride back to Rutherford on the Dugan Bakery wagon that was pulled by a horse.

Finally, my Mother was able to rent half of a house on State Street in Hackensack for $15.00 a month, which made it possible for our family, all three of us, to live together again. I'm not sure where the $15.00 for the rent came from, the relief agency or possibly Uncle Harold. I do know that Uncle Harold gave my Mother money every month and helped us in so many ways.

I remember going down to the relief agency and getting food. There was a long table with paper bags full of groceries and canned goods donated by local grocery stores. Each person got one bag, plus sometimes, a bag of potatoes. When I got home with the groceries and opened it, we often still didn't know what there was because the cans were missing the labels. So, we just shook the cans. If we heard a sound, it contained liquid, maybe soup, or maybe fruit cocktail. If there was no sound when you shook the can, it was probably beans or stew or something else like that.

During the Depression, my mother contracted tuberculosis. I came home from grade school one day to see an ambulance in front of our house and my mother being brought out on a stretcher. She was taken to Bergen Pines, a TB sanitarium in New Jersey. My brother Doug and I were sent to live with our aunts again after that. The repeated moves we went through resulted in school changes and demotions for the two of us quite often in our early years.

When my mother was finally released from the sanitarium about a year later, we were able to get back together as a family. I was in the fifth grade as a result of all the demotions from changing schools. The State Street School, which I was attending at the time, thought I should be in a class one grade level higher. They had me skip the sixth grade and jumped me to the 7th grade. This is when all my school problems began. Basic English, grammar, and math were taught in the sixth grade. I missed all that and was just never able to catch up.

I remember we lived at 403 State Street in Hackensack, New Jersey, in a very old house. Our neighbors were the Swinton sisters. They were four old ladies who lived in the house next door. Their house was even older than the house we lived in. One of the sisters, the oldest, was an artist who spent much of her time painting. Being the Depression at this time, I also had to work. Times were tough then and we used every means at hand to save what little money we had; so did everyone else. So, when Miss Swinton offered me a job, I accepted. It was only a winter job but I did earn 25¢ an hour doing it.

I would go down in the ladies' cellar and pick out all the unburned coal from the scuttle and then throw the ashes away. With a coal furnace you have to shake it down to remove all the loose ashes. The furnace had a built in shaker. After shaking the fire down, the ashes fall to the bottom so they can be removed and placed in a scuttle bucket. The fire in the furnace was continuously burning, so after shaking it down all you had to do was add more coal to the top of the fire. This provided heat throughout the house.

I would spend maybe ten or fifteen minutes each day picking the unburned coal out of the ashes so it could be used again. I would reach into the scuttle bucket which held the cold ashes and feel around for lumps of unburned coal. These I would put back in the coal bin so they could be burned in the furnace later.

The coal burned in the furnaces at that time in the East was hard coal, anthracite, rather than soft bituminous coal. I can still remember that old lady writing down the number of minutes I spent each day in her little book. When my total time worked came to sixty minutes, she would hand me 25¢. That 25¢ could buy five loaves of day old bread at a little bakery across the street from where we lived.

I had a job the summer before my junior year in high school with a milk delivery man. He would pick me up at my house at midnight. We would drive to the dairy, load up the truck, and be back on the road by 1 A.M.

In those days, milk came in quart glass bottles. I had two metal carriers that held six bottles of milk in each. My job was to run to the house and place the bottles in the milk box. When a customer lived in an apartment, I would have to run up and down the steps in the building dropping the milk at the apartment doors. Although the driver would also help delivering the milk, he made me deliver to all the apartments. That meant running up and down three or four flights of steps.

Our deliveries would take us until six o'clock in the morning when the delivery man would drop me off at my house so I could have breakfast. He went back to the dairy to load up the truck again for his commercial customers. He would pick me up again around eight or nine in the morning and we would make deliveries to stores and restaurants.

After the first night on the job, I was in such pain I couldn't get off the couch. My mother used liniment to try to ease the pain. I spent the rest of the day, and night, in bed. That first night was the worst. After that, my muscles got used to the work and I was fine. I got paid a dollar a night! In those days that was great money and I was sure glad to get it. One of the things I will always remember about that job was the delivery man himself. He smoked the worst smelling cigars, one after the other, all night long.

Prior to the turn of the century, most Americans lived and worked in an agricultural society, supporting themselves and their loved ones on the family farm. It was the family and extended family that took it upon themselves to care for their own—those who might be too old to work or were unable to provide for themselves due to disability caused by injury or illness. As the Industrial Revolution chipped away at the security of the agricultural community, however, this trend started to reverse itself. The family as a self-supporting economic unit became much less prevalent in American society. Tens of thousands of individuals and families left their farms and moved to the nation's industrial centers. Urban populations exploded. Shortly thereafter, the Great Depression again changed the face of America sweeping better than 25% of the nation's working class into unemployment and to the brink of starvation. It was within this environment of economic upheaval and social instability that the Social Security Act of 1935 emerged to insure the income of those workers who had reached a retirement age of 65.

The Social Security Act of 1935, having passed both Houses of Congress, was signed into law on August 17, 1935, by President Franklin Delano Roosevelt. One of the first requisites of Social Security proved to be of monumental proportions; register all employers and workers throughout the country over the next two years. Starting in January, 1937, workers would begin to accumulate credit toward old age benefits. The initial registration initiatives were managed and carried out by the U.S. Postal Service since, at that time, there were no Social Security employees like there are today. It was the postal workers who collected all of the completed registration forms and hand typed the Social Security cards, including the individual's assigned Social Security number. Local postal workers then returned the cards to the applicants in their area. Paperwork was forwarded to the main Social Security office in Baltimore where all the numbers were registered and employment records were originated and maintained. In the first year of registration alone, from 1936 to 1937, Postal workers handled over 35 million new Social Security cards.

Social Security has been providing payments to the retired, disabled, and orphaned for 58 years, ever since the first payment made in 1942. Bob Michel's job with the Bergen Evening Record delivering newspapers after school thus made him one of Social Security's first registrants. Although a program that has become second nature to us all, it is also one that has become the poster child for controversy as the federal government continues to find ways to fund its monumental annual budget.

Robert Michel, PhM3/c USN Malaria Control Unit Cactus. Another job I had was delivering around 200 newspapers every afternoon, The Bergen Evening Record, for the princely sum of 25¢ per day. It was during this time that Social Security came into being, so I had a penny or two taken out of my pay for that.

I also worked as an attendant in a parking lot where we charged ten cents per car and earned a penny for each car parked; anything to earn some money. Later, I worked in a local movie house, The Fox Theater, as an usher from four in the afternoon to midnight, five days a week for just $13.00 while still trying to go to high school. There was no such thing as child labor laws or minimum wage in those days.

In 1938, or maybe it was '39, when I was still in high school, I made a friend whose father was the janitor of an apartment house down on the corner. His name was Bob Kaufoldt (COW-fault). His mother and father were of German descent and were supporters of Hitler, so Bob joined the German-American Bund.

Bob was forever telling me wild stories about what took place at their camp somewhere in New Jersey in an effort to get me to join. I will tell you, from the stories he told me, it was tempting. Fortunately, though, I did not join. This guy had an IQ somewhere around 70, so he quit school after the eighth or ninth grade. I lost contact with him and have never seen or heard from him since. I do remember that he changed his last name (not legally) to Heidi. I imagine he was getting some negative feedback with a name like Kaufoldt after Hitler invaded Poland.

At one time, though, Bob and I were good friends and did a lot of things together. I remember his mom and dad and Bob and I went to some kind of German picnic where they had a huge caldron of Manhattan clam chowder cooking. I don't even want to imagine what kind of get-together that was all about. On Sundays, Bob and I would walk up Route 4 to the Washington Bridge just to look at it. It was about four miles from our house to the bridge. When we walked home there was always somebody selling clams along the roadway for 25¢ a dozen. If we had twenty-five cents between us we would buy a dozen and eat them on the way home. Clams I liked, but raw oysters were a problem with me.

My junior year in school was probably my happiest in some ways, and unhappiest in other ways. I was a popular guy in school and was elected King of the class along with the Queen, a girl named Margie Spreen. I thought she was the most beautiful girl I had ever known. I would go to her house sometimes and her mother would serve us cookies and milk and Margie would try to teach me how to dance. I never did learn very well and never was a good dancer. Margie was a fantastic dancer. Her family had money and sent her to dancing classes for ballet and tap. Sometimes she would perform in the auditorium at school and I would watch her with tremendous pride because I thought of her as mine. Of course she wasn't. I think her mother considered me from the other side of the tracks. They lived in a large house in the better section of town and her father owned a business. I guess she was right. I didn't have the manners or friends or money that Margie had. Margie was the happy time of that year.

The unhappy time was a result of not being on the football team. The coach thought I should play on the team or at least try out. I couldn't because I had to work and couldn't attend practice. He never accepted that and disliked and demeaned me even in my sophomore year for the same reason. Twenty five years later, at the only school reunion I attended, he publicly referred to me as "Bobby" in a way that was insulting.

During my two years in high school I just didn't have time to study or play football or anything else and I kept falling asleep in class. In June of 1941, having just completed my junior year of High School and flunking, I decided to join the regular Navy and make it my career. I was just 17 going on 18 that June and I needed my mother's signature because of my age to join up. At the time, the Navy offered what they called a skivvy cruise that lasted from 17 to 21 years of age. That's how I ended up in the regular Navy. As I recall, I enlisted in Patterson, New Jersey, in June of 1941. Once I enlisted, I was sent to New York City for processing and from there I was sent to Boot Camp in Newport, Rhode Island.

Built snug up against the shores of Narragansett Bay lies the city of Newport, Rhode Island, known since 1775 as the "Cradle of the American Navy." For 225 years, Newport has enjoyed a special relationship with the Navy, its shipyards providing John Paul Jones with his first command, home for the U.S. Naval Academy during the Civil War, and the site of future President John F. Kennedy's wartime training for duty on PT Boats.

When 17-year-old Bob Michel enthusiastically marched through the gates of Naval Training Station Newport in June 1941 to embark on his new career, the city itself was about to enter a period of unprecedented growth. A four-year surge that lasted throughout the war years swelled the population of the city and its surrounds by over 162,000 military personnel alone. During the war, over one hundred naval vessels called Newport, Rhode Island, their homeport.

Robert Michel PhM3/c, USN Malaria Control Unit Cactus. When our group arrived at Boot Camp in Newport, we received all our clothing and equipment that we stencilled with our names in white ink. They shaved our heads, issued us the '03 Springfield rifle, and started us on exercises, marching, and all kinds of drills. The exercises were usually with our rifles. After the first day, most of us could hardly walk or do much of anything because of the pain in muscles we had never used before. After the first day, I felt like I did the first day on the job I had in high school delivering milk, only this time, my arms were also sore from the shots.

Boot Camp was also the time for all the shots in our arms, tetanus and others. I remember that many of us had big reactions to the shots we received. We must have received three or four shots, all at the same time. That first week was hell.

Any man who has gone through military basic training remembers well vaccination day, in the parlance of recruits, "shot day." Lining up in single file and marching into that special place hidden away in the training station where every vaccine deemed by the brass to be necessary for a future seaman's health was administered.

With typical military efficiency, corpsmen were lined up in two neat rows reminiscent of the gauntlet devised by early American Indians to test the courage of their captives. On the order to move forward from the Chief, recruits were stabbed, jabbed, and otherwise assaulted on both arms by sadistic Corpsmen armed with glass syringes. Reaching the end of the line, recruits were left to wipe the blood off their arms, wincing in shock. The next day brought the full realization of the wounds inflicted when it was discovered that both arms were swollen and red. Arms throbbing in agony when a victim's tolerance for pain was tested by a well-placed punch to the arm from a grinning buddy.

Robert Michel PhM3/c, USN Malaria Control Unit Cactus. When we went to the firing range, we used the '03s. We were never given pistols and never qualified with them. I don't think I ever did well firing the old '03. I think the best I did was hit the target. It didn't really matter because most of our time was spent on the parade ground learning how to march, lots of close order drill, with and without the rifle. We spent a lot of time on calisthenics both with and without the rifle. I think I still remember how to do the left shoulder, right shoulder, port, present arms. I don't remember spending much time in a classroom except when the Chief Boatswains Mate showed us how to tie knots.

And the gas house! That I really remember. It was not a pleasant experience. The gas house was this little house where you had to wear a gas mask and they pumped in tear gas. I haven't thought of that since 1941—a long time ago.

I don't recall exactly how long boot camp was, but I think it lasted two months or so. Our designation at that time was Apprentice Seaman. After we graduated and got out of Boot Camp, we were promoted to Seaman 2nd Class. With the promotion, we went from $21.00 a month to $34.00 a month. I was the sole support of my mother and younger brother, Doug, at the time, so I was allotting all of my pay except $5.00 a month to them at home. Even later when I was on the Canal, all but about $5.00 a month went home to the family. I was what you would call one "always broke" swabby.

During Boot Camp, we always wore our whites, not the dungarees we later wore. The shoes were regular Navy issue black oxfords and probably the most comfortable shoes I have ever had. Much later on the Canal, we were issued "boondockers." They were a Marine shoe made of a rough, unfinished leather that couldn't be shined.

In boot camp we didn't have a locker to keep our clothes in. Every stitch of clothes we owned had to be rolled up right and placed just so in a sea bag that was made of white canvas. We kept our toiletries and other personal gear in a little bag called a ditty bag that went into the sea bag as well. When we had an inspection, we had to remove everything from our sea bag and lay it out on our poncho on the deck in a certain way so that the clues all faced in the same direction and the square knots were on top. After inspection, back into the sea bag everything went with the pea coat on the very bottom.

I should explain what clues are since, today, they are no longer used in the Navy. Clues were the strings that were used to tie up your uniforms once they were rolled up. Each piece of clothing was required to be rolled up in a special way so they would all fit in the sea bag and looked pressed when they were worn. Clues were also the lines that went from your hammock to the steel ring that was used to fasten your hammock onto a wall or pipe or wherever you needed to hang it. Once both ends were hung up and fastened you could get in your hammock and use it. Most of us fell out the first few times we tried to get into it. There is a trick to getting in and out of a hammock.

We were issued a hammock to sleep in; no metal racks, no beds at all. The hammock was issued to us at the same time that our sea bags and clothing were issued. It included a thin mattress and a fairly long manila line. When you closed up the hammock, it meant tying a series of half hitches along the hammock with the manila line until it resembled a long tube of canvas. The clues and rings were tucked inside before you tied it up. You then wrapped the hammock over the sea bag and down the sides and tied the whole mess together. You then carried the whole kit over one shoulder. Everything you owned was now right there on your shoulder. If you were transferred, no matter where, you took that sea bag and hammock with you.

There was a big Hollywood movie popular a few years back starring Steve McQueen, called "The Sand Pebbles." In the movie, Steve McQueen plays the part of a sailor aboard a gunboat in China and is shown on his way to his new ship carrying his sea bag and hammock over his shoulder.

Both our hammocks and sea bags were made of a white canvas and what a job it was to keep them clean. We had to constantly wash them using a scrub brush and brown soap.

We also stood fire watches, as I recall. The watches lasted four hours and we stood at parade rest in our whites with the leggings all laced up and holding our rifles.

Fire watches were, for the most part, a recruit's introduction to things military. A recruit was required to take his turn manning a guard post at a strategic location in a barracks or other military building. The principal duty of the recruit was to be on the alert for fire and, in the event, to pass the warning in a militarily prescribed manner. Fire watches also served as a time for the recruit to learn the Eleven General Orders and to familiarize himself with the military procedure of guard duty and his responsibility to the rest of the unit. The importance of obeying a superior's orders unquestioningly and learning proper military procedures were well served by the ritual of fire watches.

Robert Michel, PhM3/c USN Malaria Control Unit Cactus. Unlike the Marines, we didn't have a DI or drill instructor in the Navy. We had a Chief Boatswain's Mate. He was the boss and was always referred to only as "Sir." He was an old timer with lots of time in and lots of hash marks. The hash marks I think were gold. Gold hash marks designated four years of service with good conduct for each stripe. Red or blue hash marks merely denoted four years of service for each one.

While in Boot Camp we were given our first liberty and a guy I met there, Eugene "Scotty" MacConnell from Kingston, New York and I went into Newport and had a lobster dinner. I made contact with Scotty about a year or two ago. He is now a preacher in Ontario, New York. He was the one who persuaded me to "strike" for Hospital Corpsman after Boot Camp instead of pursuing, what I thought at the time I wanted to be, a Radioman. As it turned out, I made the right choice.

Scotty and I went on to Brooklyn Naval Hospital for Corpsman training. I don't think Scotty ever went overseas throughout the entire war but he did eventually make Chief.

At first, while we were in Brooklyn, Scotty was my best friend. He and I would hitch hike up to his home in Kingston, New York, on weekends. His mother, younger sister and brother were his entire family. I don't recall whatever happened to his father. His mom would offer us a shot of Scotch after we arrived and I felt like a real grown-up person.

It was on these weekend excursions that Scotty introduced me to a beautiful girl named Marion Britt. I would see her every time we went to Kingston. I remember sitting in the front seat of a car with her at night necking and listening to "I Don't Want To Set The World On Fire" by the Ink Spots. That became "our song." I was just eighteen and in love. I tracked her down about five years ago; she of course, was married with grandchildren. After talking a couple minutes she asked me, "Did you marry Martha?" I told her that didn't work out and after a few more minutes we hung up. I will tell you about Martha later on.

Early in 1941, Congress approved and provided funds for construction of a two-ocean Navy. With world events spinning out of control in Europe and the Pacific, leading American strategists not only saw war as imminent, but also determined that it would not be until 1943 that U.S. war production would actually be able to kick into high gear. Given the green light to proceed and the money to spend, the United States Navy started an expansion program that would see the fleet grow to literally thousands of ships over the next four years. As well, the U.S. Navy Hospital Corps would grow exponentially from a mere 4,000 servicemen in 1941 to over 132,000 personnel by the end of the war, meeting the needs of casualties in all theaters providing critical transportation, treatment, and convalescence. Education and training of new sailors as Hospital Corpsmen required the only active Hospital Corps Training School located at Portsmouth, Virginia, be expanded as rapidly as the fleet. Consequently, a temporary Hospital Corps Training School was established at the old Brooklyn Naval Hospital in 1941, adding to the capabilities of Portsmouth. Though the hospital itself had served the fleet well over the years, Bob Michel was enrolled in one of the very first classes offered by the only Hospital Corps Training School at a location other than Portsmouth, Virginia.

Robert Michel PhM3/c, USN Malaria Control Unit Cactus. Brooklyn Naval Hospital was a very old hospital located next to the Navy Yard. Even though it was old, it did have good accommodations for the petty officers in the way of private rooms. When I first went there for Corpsman Training in 1941, however, I lived in a barracks of some sort. It wasn't bad, the food was good and so was the liberty.

Fresh out of Boot Camp in September, 1941, and on his way to his new duty station at the old Brooklyn Navy Hospital for medical training, things were going well for Bob Michel. He had the security of a steady job, regular paychecks which he sent home to his mother and brother, a roof over his head, and the prospect of an education that would serve him well in years to come. With the shadows of the Depression receding, one couldn't ask much more from life. The growing conflict in Europe and escalating confrontation in the Pacific were matters of no consequence to one with prospects as bright as Bob. Little did he know, however, that the clouds of war were rapidly changing from shades of gray to dark black as he reported for training in New York.

As early as September 1, Japanese merchant ships plying their trade all over the world had begun receiving their initial movement orders from the Japanese Imperial Navy; one of the first signs that Japan was already implementing their war plans. Concurrent with the militarization of its merchant fleet, many of the naval and air units assigned to the occupation of China were being recalled to the Japanese homeland for refitting and reorganization. Japanese military units embarked on an aggressive schedule of practice landings and beach invasions, honing the deadly skills of its newly organized expeditionary forces. It was now only a matter of time and incident before the entire world erupted in all out war and Bob would be swept into its vortex.

Robert Michel, PhM3/c, USN Malaria Control Unit Cactus. It was there at Corpsman School that I first met Dave Minton. Before meeting Dave, I was hanging around with Scotty MacConnell. The more I got to know Dave, the more I started going places with him and not so much with Scotty. It was there at Brooklyn Naval Hospital that Dave and I formed a close relationship that would last for many years.

Dave and I would go on liberty together into Manhattan, stop at one of our favorite bars at the time, I think it was on E. 51st St., and drink Manhattans till we couldn't see straight. We would head over to Times Square for a while and then we would get back on the subway and head over to Brooklyn and either take the trolley back to the hospital or walk back along the famous Sands Street to the hospital. I guess I should say the infamous Sands Street. Sands Street was the locale of many, many tailor shops, bars, and tattoo parlors. Everyone who went to the Navy Yard or Hospital either walked by it or rode by it on a trolley car. It was lined with tailor shops where Dave and I bought our very first "tailor-made" dress blues. Lord, they were beautiful.

The dress blues were made of a gabardine material instead of the regular wool and were satin lined on the inside. The lining was embroidered with the design of the particular tailor. I suppose we could have specified a design had we wanted, but all we did was walk in, get measured, pay the deposit and pick up our new dress blues a few days later. Boy, was that first set of blues something else; bell bottoms, lining on the inside, the whole nine yards.

By this time we had gone from Seaman 2nd Class to Seaman First Class with three white stripes on our cuffs and, of course, another raise in pay to, I think, $54.00 a month. But, even with the pay raise, I was only keeping $5.00 on payday, so most of the time, I would run out of money before the next payday and Dave would end up paying for me when we went somewhere. Even today, I feel a great debt to Dave for all of the things that I was able to do because of him.

After Corpsman school was finished, Dave was stationed at the Naval Hospital in Brooklyn and I was transferred to the Naval Hospital at the Marine Base located at Quantico, Virginia, not too far from Washington D.C. At that time, Quantico was overrun with Marines. The town of Quantico was mostly populated by civilians who operated the stores and the bars. Many of them also worked at the Marine base and the hospital.

The U.S. Army had always provided all of its own support and service troops including medical personnel and chaplains—not so the Marine Corps. Originally formed as landing parties and shipboard riflemen, Marines have always been a part of the Navy and, as such, it was the Navy that assigned its medical and religious personnel to support Marine units. To hear Marines tell it, though, as the purest combat force ever created, they really didn't have a place for medical and religious personnel. Marines are fighters—not doctors, nurses, or preachers.

Regardless, it was in 1898, when the Marine Corps Expeditionary Battalion landed at Guatanamo Bay, Cuba, during the Spanish American War, that the Navy first officially assigned members of its Hospital Corps to field duty with fighting Marines. Ever since then, Navy Doctors, Pharmacist Mates, Corpsmen, and Chaplains have been an integral part of the Marine Corps' organizational chart.

Throughout the Second World War, Navy personnel served on every front with Marines. It was the Navy that delivered the Marines to the battlefield and it was the Navy that picked them up when the fighting was over. Through it all, Navy Corpsmen hit every beach Marines ever landed on, more often than not with the first wave in, dodging the same bullets, ducking the same shrapnel, and jumping into the same foxholes. Whenever and wherever a Marine might shout, “Corpsman up,” Navy Corpsmen were there to pull a Marine out of trouble. It was the Navy doctors, Pharmacist Mates, and Corpsmen that patched up wounded Marines and tended their sick in stateside hospitals, on ships, and in Advance Bases and Mobile Field Hospitals all over the world.

When Bob Michel left Brooklyn Naval Hospital for the Marine base at Quantico, Virginia, he would not only be living and working side by side with Marines, he would effectively be under their command. For a young man who joined the Navy because he wanted to be a sailor, Bob was about to start a whole new career with the ground pounders that would last over two and a half years and eventually take him to some of the most dangerous and inhospitable places in the world.

Robert Michel, PhM3/c, USN Malaria Control Unit Cactus. In 1941, before the war started, Marines and sailors didn’t get along too well—service rivalry, whatever. I remember it was not uncommon for fights to break out in the local bars between the Marines and sailors. I also remember the popular beer at that time was Pabst Blue Ribbon and I must say I did get to drink my share of it. That’s about all there was to do in Quantico, go to bars and drink and maybe, if you were lucky, make out with one of the locals.

There was a railroad station in town where we would catch the train to Washington D.C. when we had liberty. In 1941, D.C. really was some liberty town with lots and lots of women who were working there at the time. When I had a weekend liberty, that’s usually where I ended up. I remember that the last train back to Quantico usually left D.C. for Quantico at some ungodly early hour in the morning. If you missed that one, you ended up AWOL and would be in deep trouble.

I worked in the wards first and then as a striker for Lab Tech, X-Ray Tech, and finally, Operating Room Tech. I never really qualified for any of them except Lab Tech.

When I first arrived at Quantico, I was assigned general duty in the medical wards. Being a Marine base, most of the patients obviously were Marines. Some of them were just goldbrickers with CAT fever. Catarrh fever, undiagnosed, was what the charts usually read. Anyway, one of the nurses, a Lt.jg, who worked on that ward was a pretty regular kind of officer. She liked to play games and practical jokes with the patients and one time had me involved in one of her pranks. I think it was her best one.

She went over to a Marine lying in his bed who was a real goldbricker and told him to turn over so she could give him a back rub. He rolled over on his stomach and she pulled his PJ bottoms down below his waist and proceeded to give him a back rub with alcohol. Each time her hands came down to the small of his back, they went a little bit lower. The Marine was obviously becoming a bit aroused with all this particular kind of attention. Meanwhile, half the ward was now paying close attention to this Marine with his head buried in his pillow and with the nurse "playing around" with him.

As was previously planned, when the nurse's hands came up around the Marine's shoulders, she lifted her hands off his body and my hands came back down. It was now me giving the Marine the massage, only he didn't know it. My hands went farther and farther down until my thumb was just rubbing the inside of his thigh. I had quietly exchanged places with the nurse so that I was now the one sitting on the bed. By this time there must have been ten guys standing around watching the show.

When I thought the time was right, I poured the bottle of alcohol on his back so that it ran down the crack of his ass. It burned like hell and when he whirled around I took off like a rocket down the ward to the end of the room and ran out the door. The outraged

Marine was chasing me while trying to pull up his PJs at the same time to cover his embarrassment.

He could have killed me if he had caught me. It was wild. I think he was discharged from the ward later that day because I don't remember seeing him again. I have often wondered if he was the same Marine I ran into in September of 1942 on the Canal who kicked the shit out of me.

My next assignment at the hospital was to strike for a rating as an Operating Room Tech and the effort ended in disaster. On my first day in the OR I was to be strictly an observer and was told to remain back about two feet from the operating table. I think the surgeon was performing an appendectomy. He started out by making the incision and I started to feel a little funny. When he pushed his hand down inside the incision, I went over like a falling tree right onto my back on the white tile floor. I was dragged out into the dirty room until I regained consciousness and then they had me go back in again. By that time, the patient was being sewn back up.

The OR people expected the new trainees to be affected this way so they next sent me to the morgue to view an autopsy. Again, I was told to stand about two feet back from the table on which there was a rather obese deceased male. The doctor made his "Y" incision from the clavicle to the sternum and then one long cut down the chest and stomach to about the umbilical. As the doctor made that long incision, the fat just parted behind the blade and down I went again; once again on a white tile floor. This time I was left to lie there until I came to. The Navy and I both decided that the OR was not for me so my next move was to the laboratory.

Many of the officers stationed at Quantico had their families with them and lived in small houses supplied by the military. Because the hospital was the only medical facility in Quantico, if a family member needed medical attention, he or she could go to the hospital for treatment or they could take advantage of the outpatient services the hospital offered.

When I was working in the Lab, one of my duties was to treat patients in the outpatient program. When I would report to work in the morning, I would pick up a tray and fill it with the items needed to treat some of these patients at their homes. Most of the treatments were for blood samples or shots. In the Navy a Corpsman is authorized to do "needle work," including IVs. And if he is attached to a Marine outfit or on "independent duty," he even does minor surgery like removing cysts, and in some cases, major surgery like the guy on the sub who did the appendectomy.

Anyway, one morning, I left the hospital to make my rounds, one of which was to give a Mrs. So-and-So a shot of B12. In the Navy, the B12 was administered by injection in the buttocks with the patient lying flat on his or her stomach with the toes pointed inward to relax the gluteal muscles. One of the techniques that was taught by the Navy was to slap the buttocks lightly three times with the fourth contact being the needle itself.

So anyway, I knocked on Mrs. So-and-So's door and wouldn't you know it, a beautiful blond about twenty years old or so answered the door. I told her who I was and she had me come in. She sat down at a table and offered me her arm for the shot. I told her that the shot was administered in a different location. She was a little upset that a nurse or a doctor would not be the one to give her the shot but I convinced her that this was the way the Navy did things. I told her I had administered thousands of similar injections to other women on the base.

I convinced her to go to the bedroom and lie down on the bed. I told her to call me when she was ready for the shot, which she did. I sat down on the edge of the bed. She had her head buried in her arms. I rubbed her behind with alcohol and then slapped her butt the customary three times. Before I was able to stick the needle in, she whirled around and angrily demanded, "What are you doing?"

I'll tell you, this eighteen-year-old Corpsman was by now really nervous. Perspiration was running down my face. I wanted to be out of there. I explained the routine to her and she resumed her position on the bed. I just shoved that needle in as fast as I could, no slaps, no nothing and got the hell out of there. If she ever got another B12 shot, I don't know. If she did, it wasn't from me.

Another Quantico story. This time, I was working in the hospital lab. Quantico, as you may know, was also the training base for new FBI agents. When a new agent came to Quantico, one of their first stops was the hospital for their physical exams, which included X-rays and standard lab work. I'll tell you some of those guys were really obnoxious. It seemed they thought they were more than a few cuts above the average guy. When the X-ray techs spotted one of the new FBI guys who was a real pain in the ass they would call down to the lab and give us the guy's name, alerting us so we could do the "treatment" on him.

The head of the lab, at the time, was a guy named Bertalini who was a PhM1/c. He would go to the door of the lab and call down the line of FBI recruits for the guy whose name had been called down to us by X-ray. When "Jones," or whatever the recruit's name was, came through the door of the lab, the first thing he saw was me standing there in a white smock holding a large syringe with a spinal tap needle on the end of it. The syringe was filled with denatured alcohol, which, in the Navy, was a bright red. I am just nudging out a few drops from the tip of the needle.

Bertalini yells at the recruit, "OK, drop your pants, bend over, and spread your cheeks!" Now this is a command a recruit has heard before while going through physicals. The guy does what he is told but is in a state of absolute terror after seeing me with that large syringe. You can imagine what is going on in his mind. Bert has, behind his back, a little 10cc syringe and needle in his hand and as I step up to this poor wreck, Bert just barely touches his behind with the end of his needle. Down the guy goes, out like a light. I am sitting here laughing to myself at the memory.

I was still stationed at the Naval Hospital in Quantico, Virginia, when our war began on December 7, 1941. I was in the TB sanitarium and I remember one of the patients had a radio on. That's how I heard the announcement about Pearl Harbor. Most of us had no idea what it meant or where Pearl Harbor was.

At 7:55 A.M. on December 7, 1941, Japanese dive bombers swarmed over the Army Air Base, Hickam Field, and the naval air station on Ford Island. A few minutes earlier the Japanese had struck the naval air station at Kaneohe Bay. Bare seconds later enemy torpedo planes and dive bombers swung in from various sectors to concentrate their attack on the heavy ships at Pearl Harbor. The enemy attack, aided by the element of surprise and based on exact information, was very successful.

Torpedo planes, assisted effectively by dive bombers, constituted the major threat of the first phase of the Japanese attack, lasting

November 5, 1941
Japanese Imperial Headquarters
Navy Section, Order No.1, Tokyo

To Commander in Chief Yamamoto of the Combined Fleet:

1. In the interest of self defense and survival, the Empire is due to open hostilities with the United States, Britain, and Holland in the first ten days of December. Preparations are to be completed for the various operations involved.
2. The Commander in Chief of the Combined Fleet is to carry out preparations for the operations under his command.
3. Details will be as laid down by the Chief of the Naval General Staff.

approximately a half-hour. Twenty one torpedo planes made 4 attacks, and 30 dive bombers came in in 8 waves during this period. Fifteen horizontal bombers also participated in this phase of the raid.

Although the Japanese launched their initial attack as a surprise, battleship ready machine guns opened fire at once and were progressively augmented by the remaining antiaircraft batteries as all hands promptly were called to general quarters. Machine guns brought down two and damaged others of the first wave of torpedo planes. Practically all battleship antiaircraft batteries were firing within five minutes; cruisers, within an average time of 4 minutes, and destroyers opening up machine guns almost immediately, average 7 minutes in bringing all antiaircraft guns into action.

From 8:25 to 8:40 A.M. there was a comparative lull in the raid although air activity continued with sporadic attack by dive and horizontal bombers. This respite was terminated by the appearance of horizontal bombers which crossed and recrossed their targets from various directions and caused serious damage. While the horizontal bombers were continuing their raids, Japanese dive bombers reappeared, probably being the same ones that had participated in earlier attacks in this phase, lasting about a half-hour, was devoted largely to strafing. All enemy aircraft retired by 9:45 A.M.

Prior to the Japanese attack 202 U.S. naval aircraft of all types on the island of Oahu were in flying condition, but 150 of these were permanently or temporarily disabled by the enemy's concentrated assault, most of them in the first few minutes of the raid. Of the 52 remaining naval aircraft, 38 took to the air on December 7, 1941, the other 14 being ready too late in the day or being blocked from takeoff positions. Of necessity, therefore, the Navy was compelled to depend on antiaircraft fire for its primary defensive weapon, and this condition exposed the fleet to continuous air attack. By coincidence, 18 scout bombing planes from a U.S. aircraft carrier en route arrived at Pearl Harbor during the raid. These are included in the foregoing figures. Four of these scout bombers were shot down, 13 of the remaining 14 were taking off again in search of the enemy. Seven patrol planes were in the air when the attack started.

The enemy lost 28 aircraft due to Navy action. In addition, three submarines, of 45 tons each, were accounted for.

Bill McLaughlin, PFC, 221st Field Artillery Battalion, Americal Division. In 1936, at the age of 16, I went to the Citizens' Military Training Camps (CMTC) which, in the 1930s, were a sort of poor man's ROTC in that by putting in four years training at Regular Army camps, plus completion of correspondence courses, the training led to a Reserve Officer's commission in the Army.

The camps ran for a month in the summer, and although Fort Devens, Massachusetts, was full up, I got to go to Fort McKinley, Maine, which is out on Great Diamond Island in Portland's Casco Bay. The 6th Infantry of the 9th Infantry Division was stationed there. Other forts located nearby were Fort Prebel and Fort William.

My first year was the "basic" year and we were taught drill movements, manual of arms, etc. Since I'd already had this training with the Boston School cadets, it was a real snap for me.

At the end of the training, if we graduated, and not all of us did, we opted for a particular branch of the service for the following red, white, and blue years. I opted for the cavalry and was slated to go to Fort Ethan Allen, Vermont where a squadron of the 3rd Cavalry was located at that time.

A neighbor of mine who was a sergeant in the 110th Cavalry of the Massachusetts National Guard, hearing of my aims, convinced me to join his troop. Although I was just 16-1/2, and the minimum age for joining was 18, I was tall for my age and passed easily for 19, which I told them I was.

In 1940, three years after my training with the CMTC, my buddy Tim and I planned to join the Marine Corps. Tim was turned down as being too tall, he was 6'3", so we both reenlisted in the Cavalry again. In late 1940, our regiment was changed over to field artillery and attached to the 26th "Yankee Division." They were about to enter a year of Federal Service at the time.

We were sent down to Carolina for maneuvers for three months and we returned on December 6, 1941. The very next day, Pearl Harbor was bombed. Our Division, like all the rest in the Army at the beginning of the war, was "triangularized" from a square division and reduced from 20,000 to 15,000 men. The regiments were split into battalions and the division was composed of three infantry regiments, not the customary four.

Tom Croft BM3/c, Gun Loader Turret 1, USS Monssen DD436. In March, 1940, after I finished Boot Camp at Great Lakes in Illinois, they shipped a bunch of us to San Pedro, California by train to join up with the USS Arizona. From San Pedro, we sailed straight off to Hawaii where we stayed for a week or two before going out on maneuvers . On that first cruise of mine we went south and that is when I first crossed the equator and was initiated. I went from a pollywog to a shellback.

At the time, I was a deck hand—a Seaman 2nd Class. It would still be about a year before I made Seaman 1st Class. I stood routine watches on the ship and also watches up in the crow's nest. When we were on watches up in the towers there were always four guys up there. Other than my duties on the deck gang, I was assigned to the #1 gun turret. My station was down in the magazine and my job was to send powder bags up to the guns.

There wasn't a lot of pomp and ceremony on the Arizona. The ship was old at the time but they were always putting new kinds of guns on them for antiaircraft and stuff like that. Those old battleships were equipped good—they were just not as speedy as some of the newer ships.

When I was assigned to the Arizona, remember, it was still peace time so there was never a feeling aboard ship that the Japanese were going to be a problem like they later were. We never ran into the Japanese fleet back then out there either.

I remember we would hold training exercises out in the Hawaiian Islands. They would have a tug boat tow a target which was flying up behind it. They would tow the target across in front of us and we would shoot at it. We would fire the big guns at the target which was about 10 miles away or so. I got to see the target a few times but my job was down in the magazine loading up powder bags to send up to the guns to fire.

Being stationed in Hawaii before the war was real good duty. I liked it out there—it was real nice.

In September 1940, the Arizona cruised up to Washington state. This is when a bunch of us were transferred over to the Monssen, a new destroyer just being commissioned.

When the war broke out December 7, 1941, we were on the Monssen up in Iceland. We had taken the Hornet out on her shakedown cruise with the USS Missouri, New Mexico, Idaho and four destroyers. They had returned to the States already for transport duty between Iceland and Greenland. We stayed on in Iceland.

We knew that Pearl had been attacked and we gave those Japs a real good cussing—dirty and sneaky as they are. I had a lot of friends that were still on the Arizona when she was hit—Martin and Rose—I can still name a lot of them. It must have had a real effect on the American guys out there. Later, we really liked shooting them planes of theirs down. Shooting at their planes was better than shooting at their ships.

With the start of the war, we were immediately ordered back to the States and a task force was put together in New York with four destroyers and two heavy cruisers. We left New York and met up with the Hornet again and five troop ships and a tanker. The troop ships were carrying reinforcements to guard the Panama Canal. We went through the Canal after dropping off the troops and headed up to San Francisco. The Hornet followed us there.

NAVY DEPARTMENT COMMUNIQUÉ

No. 214, December 6, 1942
South Pacific (all dates are east longitude)

(For security reasons, released one year after the Japanese surprise attack on Pearl Harbor.)

THE JAPANESE ATTACK ON PEARL HARBOR, DECEMBER 7, 1941

On the morning of December 7, 1941, Japanese aircraft temporarily disabled every battleship and most of the aircraft in the Hawaiian arsenal. Other naval vessels, both combatant and auxiliary, were put out of action, and certain shore facilities, especially at the naval air stations, Ford Island, and Kaneohe Bay, were damaged. Most of these ships are now back with the fleet. The aircraft were all replaced within a few days, and the interference with facilities was generally limited to a matter of a few hours.

When the Japanese attacked Pearl Harbor, 2 surface ship task forces of the Pacific Fleet were carrying out assigned missions at sea, and such task forces were at their main base following extensive operations at sea. Discounting small craft, 86 ships of the Pacific Fleet were moored at Pearl Harbor. Included in this force were 8 battleships, 7 cruisers, 28 destroyers, and 5 submarines. No U.S. aircraft carriers were present.

As a result of the Japanese attack, five battleships, the Arizona, Oklahoma, California, Nevada, and West Virginia; three destroyers, the Cassin, the Shaw and

Downes, the mine layer Oglala; the target ship Utah, and a large floating drydock were either sunk or damaged so severely that they would serve no military purposes for some time. In addition, three battleships, the Pennsylvania, Maryland, and Tennessee, three cruisers, the Helena, Honolulu, and Raleigh, the seaplane tender Curtiss and the repair ship Vestal were damaged.

Of the 19 naval vessels listed as sunk or damaged, the 26-year-old battleship Arizona will be the only one permanently and totally lost. Preparations for the righting of the Oklahoma are now in process although final decision as to the wisdom of accomplishing this work at this time has not been made. The main and auxiliary machinery, approximately 50% of the value, of the Cassin and Downes were saved. The other 15 vessels have been or will be salvaged and repaired.

Eighty naval aircraft of all types have been destroyed by the enemy. In addition, the Army lost 97 planes on Hickam and Wheeler Fields. Of these, 23 were bombers, 66 were fighters, and 8 were other types.

The most serious American losses were in personnel. As a result of the raid on December 7, 1941, 2,117 officers and enlisted men of the Navy and Marine Corps were killed, 960 are still reported as missing, 876 were wounded but survived. The Army casualties were as follows: 990 officers and enlisted men were killed or later died of wounds, 396 were wounded, most of whom have now recovered and have returned to duty.

Robert Michel, PhM3/c, USN Malaria Control Unit Cactus. With all of the upheaval and excitement of Pearl Harbor, I was shortly thereafter transferred to MOB5, Mobile Hospital 5. MOB5 was a new unit that was being formed back at the Brooklyn Naval Hospital. Dave was also assigned to MOB5 so we again were reunited.

When Bob was issued orders sending him back up to Brooklyn Naval Yard to transfer him to a new mobile field hospital unit, MOB5, the race to get men and supplies into the South Pacific to block the Japanese had already begun. Bob was part of that initial rush but was by no means on one of the first ships to disembark.

On January 23, 1942, just six weeks after Pearl Harbor, Task Force 6814 had already formed up, loaded, and disembarked Brooklyn Harbor with a hodgepodge of units that were to be melded into an Army Division in the South Pacific. Passing through the Panama Canal just a week later, the Task Force headed straight on to Melbourne, Australia, arriving there on the 26th of February. After a brief layover of about ten days, the men were once again loaded onto their ships and headed northeast to Noumea, New Caledonia, their final destination.

The newly arrived units, approximately 17,000 strong, were a mixture of Regular Army units, Reserves, and National Guardsmen who were quickly reorganized by General Alexander Patch into what he called the Americal Division; an abbreviated term deriving from Americans in New Caledonia. For the most part, poorly equipped, many with weapons left over from the First World War, and a good percentage of the men inadequately trained, they quickly pulled together and later became one of the most famous fighting units of the Second World War.

The Americal Division was ordered to protect the island and defend at all costs the newly constructed airfield at Tantouta from an expected Japanese invasion. A daunting task for any occupying force given over 150 miles of beaches and a French population divided politically by the war in Europe between the Free French, those siding with the Allies, and the Vichy French, those sympathetic to the Axis cause. For the next three months, while the Americal Division protected the shores of New Caledonia, the Navy and Marines poured onto the island to build the support and supply infrastructure necessary to launch future offensive operations against the Japanese.

Bill McLaughlin, PFC, 221st Field Artillery Battalion, Americal Division. Our 1st Battalion Field Artillery was shipped out with the rest of the downsized outfits in January 1942, with Task Force 6814. Units from Illinois and North Dakota joined us on New Caledonia to become the Americal Division—Americans on New Caledonia.

Though Bob and Dave didn't know where they were headed, the first stop on their South Pacific Cruise was to be New Caledonia, stepping stone to Guadalcanal.

Robert Michel, PhM3/c, USN Malaria Control Unit Cactus. About April or May, 1942, other members of MOB5, along with Dave and I, were transported to the San Diego Destroyer Base by train on a route that zigzagged across the country. What we didn't know at the time was that our ultimate destination was to be Noumea, New Caledonia. At Noumea, our unit would be setting up the hospital. We were to ship out through the San Diego Destroyer Base. During the trip, however, we never knew where we were or where we were going. Hell, one guy swore we were even in Canada at one time because we had been traveling north for so long.

When Dave and I finally arrived in Southern California, we were assigned temporary duty during the day to keep us busy while we were waiting to ship out overseas. So, Dave and I were put on the gang that was building a new rifle range. As I recall, the range was being built below ground level that required us to dig this big, long, deep ditch.

Well, being the accomplished goldbrickers that we were, we soon figured out a way to get out of that detail. We would show up for work in the morning and then disappear, heading off to a beached ferryboat not too far from the dig. We had stumbled across it one day while we were doing a little exploring. So, instead of doing the work that was assigned to us, we would head out for the ferry and lay out on the deck in the sun all day until it was quitting time. Then we would rejoin the work detail and call it a day. Both Dave and I have fond memories of our days in the sun on the deck of that boat while the other guys were digging that ditch.

Right about the time Bob and Dave were packing up to head west with MOB5, a number of U.S. ships had already set sail for the South Pacific. Once war was declared the Navy didn't waste any time in starting a mad dash across almost 6,000 miles of ocean to set up bases in Samoa, Fiji, the New Hebrides, and New Caledonia. It would be the bases these men built that would in a very short time, accommodate the men and supplies that would soon follow. Storm clouds were indeed gathering as Japanese and Allied forces rushed headlong into one of the most wide open and obscure regions of the Pacific, both intent on controlling vital shipping and communications lanes between the United States and Australia. Whichever side won this race would eventually control the entire Pacific, from the Chinese mainland all the way to the West Coast of the United States.

One of the first ships to sail was the SS *President Monroe* in mid-March. Crammed into one of her converted holds went Thomas Gross, drafted into the Army on February 21, 1941, just one year earlier. Oddly enough, when Gross received his induction notice, his hitch with the Army was to be for just one short year. Prior to December 7, 1941, a draftee was only required to serve for a period of twelve months; with the advent of the war, however, that would soon change. When Gross sailed with the 97th Field Artillery Battalion, in what later would be called the Americal Division, his Army service had already exceeded his initial term by 16 days. Gross, like many hundreds of thousands of other young men, was now in the Army for the duration.

Thomas Gross, U.S. Army 97th Field Artillery Battalion, Americal Division. I was twenty-three years old at the time and I was in an outfit that was originally part of the 4th Field Artillery. This was a Regular Army outfit. All the men came from North Carolina, South Carolina, Georgia, Tennessee, Alabama, and Kentucky. My buddy, Oscar Currier from Maine, and I were the only ones who came from New England. Hell, I never even had grits until we were sent down to Fort Bragg.

Being Regular Army, all the sergeants had been in grade at least ten plus years. Staff Sergeant "Pop" Brower had been in the Army for over twenty-five years. When Oscar and I were made Private First Class after we had been in the outfit only four months, they almost had a revolt in the battery. The men that were the most upset were the guys that had just reenlisted for another three years when the Captain told them he would make them PFCs. I had been drafted for only one year, remember.

On 15 March 1942, our unit left San Francisco aboard the SS President Monroe and sailed under the Golden Gate Bridge into the Pacific Ocean. We had no idea where we were being sent.

The first night aboard, I went down into the hold and slept in a top bunk. When we sailed on the President Monroe, she still had a swimming pool, as I recall. Originally, the ship had accommodations for sixty passengers, that is, before the U.S. government took her over and converted her for carrying troops.

I think the ship had a 3" gun mounted on the bow and we put two of our 75mm Pack Howitzers on either side of the stern. I recall our gunners would practice shooting at "enemy subs" during the voyage; very funny, with the ship rolling from side to side the way it was. We were all alone out there and never did see a real sub the whole trip.

On about the third day out, it got so hot we all stayed up on deck, day and night. This way, we had some space to stay and sleep 24 hours a day if we wanted.

The food was so bad we paid our old mess sergeant to get us candy and shoe string potatoes from the ship's canteen. The mess sergeant had been busted down to a private, so we gave him money to get us what food he could as well.

We made one stop on that trip and that was at Bora Bora. I recall the only word we got there was that all the ladies on the island were NOT to be touched.

Bora Bora, or more correctly Pora Pora since the letter "b" does not exist in the Tahitian alphabet, can be found precisely 150 miles northwest of the island of Tahiti. Part of French Polynesia, which consists of 35 islands and 83 atolls spread out over 2.8 million square miles of the Southeast Pacific Ocean, is made up of five great archipelagos, the Marquesas, Australs, Gambiers, Tuamotus, and the Society Islands.

Bora Bora is located in the Society Islands Archipelago, which itself is made up of two groups of islands called the Windward Islands and the Leeward Islands. Bora Bora, located in the Leeward Islands, is a tiny speck of basalt hills located in the center of a lagoon. It is a mere 6.25 miles long and a total of 15 square miles in area, surrounded by a ring of smaller islands enclosed in a necklace of coral reef. Bora Bora is universally considered to be one of the most beautiful places on the face of the earth. Everything we have ever heard or imagined about a South Pacific island paradise is embodied in the reality of Bora Bora. It was James Michener who once described Bora Bora as a place "so stunning, there are really no adequate words to describe it."

Situated due south of the Hawaiian Islands and north of the New Hebrides and New Caledonia, Bora Bora's location made it a natural refueling spot for military ships heading to New Zealand and Australia. Due to its strategic position smack dab in the middle of the Allies' supplies lines, Bora Bora was occupied by U.S. troops just as soon as hostilities with Japan commenced. An airstrip was

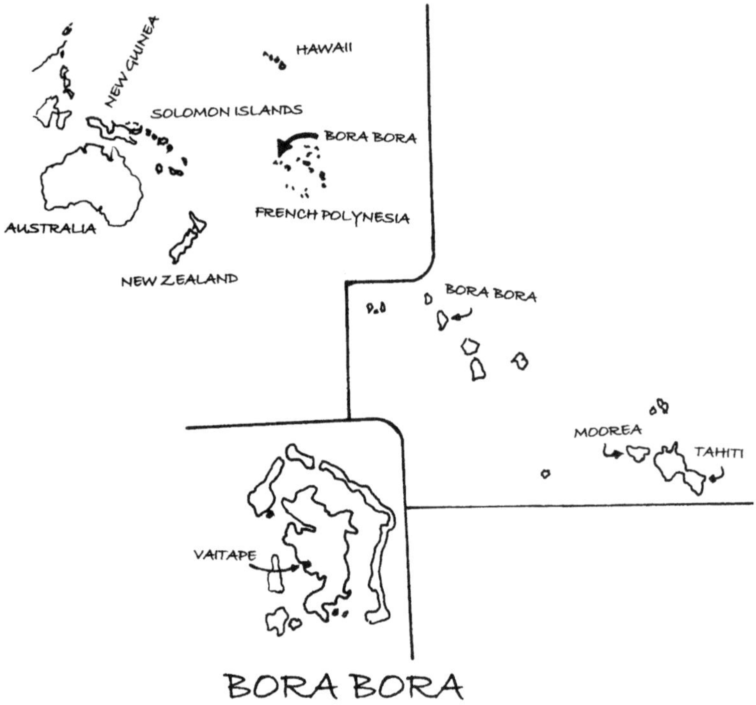

constructed, a naval refueling station was established, and the island soon bristled with scores of artillery pieces mounted along her shores to protect U.S. military assets from the threat of Japanese attack.

Today, the old World War II military airport still exists and passengers flying in are ferried to the main island in the center of the lagoon by passenger boat. A number of the old Army bunkers can still be visited and many of the coastal artillery pieces left behind at the end of the war can still be seen pointing out to sea. Bora Bora remains a popular tourist destination and the site of one of the world's most spectacular coral reefs; a breathtaking destination for divers visiting from all over the world.

The five archipelagos that make up French Polynesia, even though they fly their own flag, still remain part of the French

Republic. The French Constitution is in force on the islands and French is the official language spoken, though many of the natives maintain their fluency in Polynesian.

Thomas Gross, U.S. Army 97th Field Artillery Battalion, Americal Division. On the night of April 2nd, 1942, we pulled into this strange dock. There was one GI there who had a rifle standing guard on the dock. We asked him where in the hell we were. He said, "New Caledonia." None of us had ever heard of the place before so we all ran to where there was a map on the ship and lo and behold, there on the map was New Caledonia.

When we arrived in New Caledonia we were without our mules. We were told they would be arriving on a later transport.

The first day there we marched up what was nothing more than a bicycle track. That first night it rained like crazy; we were all in our pup tents and when we looked around the next morning, all of our gear was scattered all over the hill we were on.

We stayed up on that hill for a few days before we went farther up the island to a much neater area that had large six man tents and a great eating house which had open sides with a palm covered roof. There was also a clear stream there about six or seven feet across that looked to be about four feet deep. But you could climb ten feet up a tree, dive into that pool and never hit bottom; that water was that clear.

Finally, we got word that our mules had arrived so we headed back down to the docks at the port to get them. That was an experience I'll never forget. I got a hold of two of them by their haltered shanks and started to leave. Well, there were these natives standing there watching us. These are people we had been told were just two generations from being headhunters; tall, black, and with a huge head of orange hair. As I started moving up the dock with the mules, one of these fellows put his thumbs up to his ears and started to "whinny" just as a mule does. I heard this and I thought, God, Gross, here you are, three years away from graduating from college and here is a guy whose great grandfather was eating one of his own kind; how low have you got? Well, we just continued up the road, past the infamous "Pink House" and back to our quarters.

Tom Croft, BM3/c Gun Loader Turret 1, USS Monssen DD436. It was April 1942 when we arrived in San Francisco and we were only there about three days when, one morning, we got word we were to get under way. I remember it was real foggy that morning. I don't think anyone had seen that the carrier, the Hornet, left San Francisco with B25s on it. We left along with the Hornet. We had a complete task force and a tanker along with us.

We met up with another task force of ships, this one from Pearl Harbor and that is when we learned where we were going. Every destroyer took on fuel from the tankers. They said we were going to within 400 miles of Japan. When we got to within about 800 miles, though, we spotted a Jap fishing boat. Our cruisers sank it right away. The Hornet launched our B25s right away from 800 miles out from Japan. They all done their jobs. Some made it back to China and some didn't.

After the planes were launched we headed right back to Pearl Harbor and refueled and took on supplies. We headed right back out towards the Coral Sea.

During the first week of April, 1942, Vice Admiral Yamamoto (Commander in Chief of the Japanese Combined Fleet and the mastermind behind the attack on Pearl Harbor) wrote his good friend Niwa Michi (a geisha and confidante with whom he had corresponded for year):

> The thing that worries me slightly is that, although the war's only been going on for three months, a lot of people are feeling relieved, or saying they're grateful to Admiral Yamamoto because there hasn't been a single air raid. They're very wrong:

the fact that the enemy hasn't come is no thanks to Admiral Yamamoto, but to the enemy himself. So if they want to express gratitude to somebody, I wish they'd express it to America. If the latter really made up its mind to wade in on us, there'd be no way of defending a city like Tokyo.

The first stage of operations has been like a children's hour, and will soon be over; now comes the adults' hour, so perhaps I'd better stop dozing and bestir myself (Hiroyuki Agawa, **The Reluctant Admiral, Yamamoto and the Imperial Navy**, Kodansha International, Ltd., 1979).

Little did Yamamoto know that an American Task Force under the command of Admiral William F. "Bull" Halsey had already set sail from San Francisco. Quickly patched together with what ships were available, the little armada consisted of the carriers Hornet and Enterprise, escorted by four cruisers and eight destroyers. One of those destroyers was the scrappy little Monssen fresh off convoy duty in the Atlantic.

It had never occurred to the Japanese that the Americans would attempt to fly long range bombers off the deck of a carrier. As Tom Croft alluded earlier, Jimmy Doolittle was about to do just that. With 16 B25, twin engine, medium bombers strapped to the deck of the Hornet, his plan was to sail his bombers within flying range of the Japanese mainland and launch his planes. Once in the air, his plan called for his airmen to bomb their designated targets in Japan and to land on an airfield in China held by Chaing Kai-Shek's friendly Nationalist forces.

At 0630 on April 18, 1942, the Japanese patrol craft No. 23 Nittu Maru spotted Halsey's Task Force just 650 miles off the coast of Japan. The Nittu Maru raised the alarm by radio and was immediately sunk by the cruiser Nashville. Japanese naval authorities were not quite sure what they heard coming in over the airwaves, so they hesitated.

While the planes from the carrier Enterprise guarded the tiny flotilla, Hornet successfully launched all 16 of Doolittle's B25s.

Once the B25s were in the air, Enterprise recovered her planes and the task force turned around and headed east, making good their escape.

Doolittle and his raiding party split up into several groups, each assigned a primary target for the day's mission. Not only was the city of Tokyo bombed, but also Kawasaki, Yokusuka, Nagoya, Yokkaichi, and Kobe were all hit that day. Not a single plane was lost over Japan.

After suffering repeated defeats over the previous four months, the Raid on Tokyo lifted the spirits and raised the morale of the American people. They now truly believed that their military was on the offensive and would soon defeat the Japanese.

By mid April 1942, the U.S. Pacific Fleet was stretched to its limit, forced to protect an ocean area larger than any in the world against a far superior enemy with the capability to strike at will in any of a dozen different locations. New Zealand, Australia, New Guinea, and the tiny island bases on New Caledonia, Fiji, Samoa, and the New Hebrides all fell under the threat of imminent Japanese invasion. When naval intelligence gathered information of a large Japanese invasion fleet moving into the Coral Sea, all the pieces fell into place for a do or die showdown. The outcome of the war itself hung in the balance.

All over the South Pacific, military and civilians alike prepared for the worst. Contingency plans for evacuating towns and villages were swiftly put into motion. Army and Naval units guarding island shores started the grim task of preparing lines of retreat back into isolated, mountainous regions, making ready to fight on in small guerilla groups if need be. Civilian populations were evacuated with women and children leading the exodus from homes likely to stand in the path of oncoming Japanese landing forces.

Just one day after General Wainwright surrendered to Japanese forces on Corregidor, the American and Japanese carrier forces clashed in what would shortly be called the Battle of the Coral Sea.

The engagement that took place on the 7th and 8th of May, 1942, represented a number of firsts in the annals of warfare. The Battle of the Coral Sea was the first ever American attack launched against enemy aircraft carriers. On May 7, the first day of the battle, planes from the carriers Lexington and the Yorktown slammed into the Japanese invasion fleet. The first aircraft carrier battle was thus engaged. American fliers acquitted themselves well by seriously damaging the Japanese light carrier, Shoho. The next day, May 8, U.S. carrier planes again went on the attack, seriously damaging two more Japanese carriers, the Sikaku and the Zuikaku. Japanese planes broke through the defensive screen of the U.S. fleet and destroyed the Lexington, while seriously damaging the Yorktown.

The Battle of the Coral Sea forever removed any doubt in the minds of military planners that the outcome of the war in the Pacific would hinge upon aircraft carriers. For the first time in naval history, two opposing naval forces so far apart they couldn't see each other, battled to the death. The Battle of the Coral Sea was, quite frankly, where carrier combat really began.

The outcome of the battle was, perhaps, a tactical victory for the Japanese. They destroyed one U.S. carrier and seriously damaged another, cutting U.S. naval carrier strength in half in a single engagement. The strategic victory, however, was clearly achieved by the U.S. Navy when they knocked two of Japan's heaviest carriers out of the war and forced a Japanese withdrawal from the area, thus saving every island in the South Pacific supply chain from invasion. New Zealanders and Australians, to this day, thank the U.S. Navy for saving their countries, their families, and their way of life.

Thomas Gross, U.S. Army 97th Field Artillery Battalion, Americal Division. We never heard too much from the Japs while we were up in the hills outside Noumea and didn't really hear much about them bombing the island of New Caledonia at all. Of course, we did know a lot about where they had been and where they might be going next, so our Battalion Commander had us digging retreat trails up into the mountains so we could use them if and when the Japs landed. This all changed, of course, when the Navy met them at the Battle of the Coral Sea. God Bless the Navy for that.

Chapter 3

South Pacific Cruise

June 1942-August 1942

BEFORE WE GO too much further with Bob's story I feel it would be helpful to jump ahead in time and relate a story that was recently told to me by Robert Otto. He was just another young man born and raised in a small town in Ohio who had never traveled outside its limits until he joined the Navy in 1941. Robert was a Boatswains Mate and 20mm gunner who served aboard the destroyer Grayson for the better part of two years. During that time he had seen more of the world than he ever dreamed existed—from Iceland, to the Panama Canal, to San Francisco, to Hawaii, and beyond. His story's relevance will be appreciated as the reader progresses with this narrative. I am not going to quote Robert verbatim, but will paraphrase to make my point.

As Robert tells it, the Grayson was one of the most active destroyers operating in the South Pacific as early as the spring of 1942. The Grayson would pull into so many different islands so frequently that the crew for the most part never knew where they were or even the name of the island they just left. If they did know the name of the island, they were seldom able to pronounce it. This went on for months and months until one day, Robert was leaving his compartment somewhat in a funk and ran into an old Chief who had been in the Navy for 30 years or more. The Chief stopped Robert and asked him how he was doing, guessing correctly that he might be having somewhat of a down day. In an effort to shore up his spirits, the Chief told him, "Look at it this way, son, when this is all over and you get back home you will be a regular Lowell Thomas when you start telling the folks about all the places you've been. They won't know what you're talking about when you start in on places like Fiji, Bora Bora, Tonga, Efate, Aore, Tahiti, Noumea, Tulagi, Espiritu Santo, Vella Lavella, Kolombangara, and Gavutu. They'll just look at you and you'll knock 'em right over with the stories you'll tell them of all the places you've been that they've never even heard of. You're going to enjoy it. You'll really be something, kid."

In later years, Robert realized just how right the old Chief was, and to this day still gives himself a chuckle over that story just like he did when he told it to me. With that in mind, let's take our South Pacific cruise, shall we?

Robert Michel, PhM3/c, USN Malaria Control Unit Cactus. In June, 1942, we were put aboard a civilian ship, the President Monroe. We couldn't wait to get going, but I think we were more apprehensive than excited about it. We knew we were going overseas but didn't know where or what to expect, except that we were going to be setting up a hospital somewhere.

We figured if we were going to be setting up a hospital, it had to be for wounded men. If it was for wounded men, we knew it wouldn't be too far from where they were wounded. Excited about going overseas would be a valid description, but apprehensive would describe my feelings as well.

That was some trip. We were told this was to be the last sailing of the President Monroe under civilian control. It became obvious very quickly that the owners were saving money any way they could. We got two meals a day and they were the worst—at least they were the worst until we got to the Canal.

Tom Croft, BM3/c Gun Loader Turret 1, USS Monssen DD436. After we left the Coral Sea, we headed right back up to Pearl Harbor and loaded up the ship and headed right out for Midway Island. It was a hard battle. Both the Japs and us lost heavy. The Yorktown sank along with the destroyer Hammann.

After the battle, we headed right back to Pearl again taking on fuel, ammunition, and staples. We headed back to the Southwestern Pacific and our next stop was Guadalcanal.

Vice Admiral Yamamoto, Commander of the Japanese Combined Fleet, firmly believed that Japan could win the war if she did so during the first twelve months of hostilities. In fact, Yamamoto was overheard on many occasions saying that in the first six months after entering the war, he could run wild in the Pacific. After that, he would make no promises. He felt the sleeping giant of American industrial might would have been goaded into wakefulness by then.

True to his word, Yamamoto planned his final, decisive blow against the United States Navy to be delivered by a massed attack on Midway Island, located just 1,100 miles from Hawaii. Once Midway fell, only Hawaii would block Japanese forces from hitting the West Coast itself. By gathering over 162 of his nation's naval ships in harbors all over the Japanese islands, Yamamoto's plans called for a diversionary attack on the Aleutian Islands with his main invasion force hitting Midway itself. If he succeeded, he felt he could draw the U.S. Fleet out into the open for one last surface battle to settle once and for all the issue of supremacy of the Pacific.

Ever the gambler, Yamamoto set sail from Japanese waters on May 27, 1942.

Vice Admiral Nimitz and his Naval strategists had long felt Midway would be Yamamoto's next objective. What proved to be the final ace in Nimitz's hand was the fact that U.S. cryptologists had long since broken the top secret Japanese codes and were listening in on their most intimate conversations. Naval Intelligence confirmed what Nimitz and his staff had already deduced; Midway was the next objective of the Japanese Fleet. Not only did Intelligence provide Nimitz with the Japanese target, they handed him the date and hour of the attack and the strategy laid out for the diversionary attack scheduled to hit the Aleutians. Having seen his opponent's cards, Nimitz was free to ignore the Japanese feint to the north and could concentrate his forces just 300 miles northeast of the island of Midway and wait for Yamamoto to show up.

Ordering around-the-clock repairs to the aircraft carrier Yorktown, which was seriously damaged just three weeks earlier in the Battle of the Coral Sea, Nimitz wanted all of his available aircraft carriers ready to meet the Japanese Fleet. Ready to go were the Hornet and the Enterprise; all he needed now was a fit Yorktown, which ultimately made the rendezvous in fighting trim. On May 31, Nimitz had his forces in place waiting for the Japanese attack he knew was scheduled for June 4. Now, the final advantage would go to the fleet which sighted the enemy first.

On June 3, 1942, a single Catalina bomber on scouting patrol from the airfield on Midway spotted the Japanese Fleet bearing down on the island. B17 bombers were launched from the island's airfield and though they launched the first attack against the Japanese, not one of the bombs they dropped so much as scratched an enemy vessel.

The next morning, Admiral Nagumo, the Japanese Task Force commander, ordered over 108 planes from his four carriers (the Kaga, Kiryu, Soryu, and Akagi) to take off, armed with fragmentation bombs. Their first target was the island itself and the U.S. airfield located there. He kept his scout planes in the air diligently searching for the American Fleet which he was sure was lurking somewhere nearby.

By the time the first waves of Japanese planes returned from their attack on Midway, Nagumo had still not located a single U.S. Naval vessel. With initial reports coming in that damage inflicted on the airfield at Midway was minimal, Nagumo gambled again. He ordered all of his planes to be armed with high explosive bombs rather than armor piercing munitions (necessary in any attack against enemy surface vessels).

Right about this time, Enterprise and Hornet launched their torpedo bombers in the first wave of the U.S. attack on the Japanese carriers and support ships. In all, over 58 Navy torpedo planes from the two carriers engaged the enemy. All but six planes were destroyed by Japanese fighters or antiaircraft fire. While the surviving torpedo bombers limped back to their carriers, Enterprise and Yorktown launched their dive bombers. Enterprise flyers destroyed the Akagi and knocked the Kaga out of commission for the duration of the war. Not to be shortchanged, Yorktown's dive bombers entered the shooting gallery, hitting the Soryu. The final coup de grace for the Soryu, however, was delivered by a nearby U.S. submarine, the Nautilus, who promptly fired three torpedoes into her burning hulk, sending her to the bottom.

Now the last undamaged Japanese carrier, the Hiryu, retaliated with a vengeance when her planes came screaming down on the Yorktown, setting her ablaze. Hit just three weeks earlier in the Coral Sea, the Yorktown found herself dead in the water, this time incapable of continuing the fight. Planes from the Enterprise and Hornet quickly located the Hiryu and destroyed her, effectively deciding the battle.

The Battle of Midway cost the Japanese 4 aircraft carriers and along with them, the irreplaceable pilots that would be needed in future actions. The action effectively ended any further Japanese plans to dominate the Eastern Pacific.

On the American side, the Yorktown was lost; however, the U.S. Navy had just walked away with the most decisive victory in the history of naval warfare. The balance of naval power in the Pacific had just been evened and the Japanese were placed on the strategic defensive; a situation which was not to be reversed for the remainder of the war.

With Japanese forces forced to withdraw, the United States, for the first time since Pearl Harbor, was clearly ready to go on the offensive in earnest. It was now just a matter of time and deciding where best to strike.

Robert Michel PhM3/c, USN Malaria Control Unit Cactus. For sleeping, we were assigned cots in the hot hold of the ship. But after one night in the hold I decided to take my hammock up on deck and sling it under a boom that was used for loading and unloading. I spent the rest of the trip sleeping up there on deck under the stars. If the ship rolled or tossed the hammock never moved one way or the other.

After I slung my hammock up there, a lot of the other guys tried to do the same thing. But after a couple hours of frantic activity there wasn't any room left up there for hammocks.

It was a great trip, though. Nothing to do day or night. Lots of poker games going on at all hours. The only problem was, though, that we weren't allowed to be in our hammocks during the day. So we would find places in the ship's superstructure where we could sit or lie down without fear of falling asleep and falling overboard or down onto the steel deck. Because the food was so bad, Dave and I sneaked down to the officers' galley one day and stole two whole baked hams, along with a couple loaves of fresh baked bread. We ran like hell down to our bunk locations and proceeded to make ham sandwiches, which we took up on deck and sold to the poker players.

If we wanted to take a shower on the Monroe, we had to take it with seawater and use what they called salt water soap. It got most of the dirt and perspiration off but left a layer of salt all over your body.

On the way, we docked at American Samoa for a day, but none of us were allowed to go ashore. When we crossed the equator, there was the usual celebration for the first timers. We were duly designated "shell backs."

On the day the SS President Monroe crossed the equator, like every ship civilian or military that had preceded it for hundreds of years, it was time for those aboard to set aside thoughts of war and pay homage to the mythical King Neptune, legendary god of the Seven Seas. Crew members had probably been storing up the ship's garbage, and anything else of an equally vile nature, for days. Bob, the rest of MOB5, and all the Marines aboard were about to have some great fun; albeit at their expense.

The ceremony of initiating the "pollywogs" (those who had never crossed the equator), to the society of trusty "shellbacks" (those who had undergone the rites of passage), is one that dates back to unknown pagan religious rituals. On said date and at the prescribed time, the ship's pollywogs are led by the shellbacks onto the deck of the ship, usually forced to wear humiliating garb of some sort, often swim trunks. The pollywogs are paraded around the ship's decks in ritualistic fashion. The pollywogs are forced to have their heads shaved, and on some ships, to run a gauntlet of shellbacks who strike them with rolled up socks and towels. The pollywogs are then made to crawl through the stored garbage which has been heaped on the decks for ceremonial purposes. At the end of the gauntlet of abusive shellbacks and reeking garbage is King Neptune himself, sitting on a throne erected for the ceremony. Having suffered indignity and humiliation, the pollywogs are made to kiss the belly of King Neptune. King Neptune invariably is the man on board with the biggest beer gut of all, whose belly is covered with a vile concoction of oil, grease, and various available condiments. Once "King Neptune" has determined that a pollywog has suffered enough (twisting and writhing on his gut) the pollywog is declared a true "Shellback."

Once a pollywog undergoes the prescribed rites of passage to the satisfaction of all, he is issued an official "Shellback" identification card. Wise is the new "Shellback" who keeps handy his identification card so that the next time he crosses the equator, he will not have to undergo the initiation ceremony again.

Robert Michel PhM3/c, USN Malaria Control Unit Cactus. We were on our way to Noumea, New Caledonia, where we were to set up a mobile field hospital. I believe it was now May, 1942.

Marine Sergeant, Section Leader 75mm Pack Howitzers, I/3/11. We left San Francisco on the SS Ericsson. You know what they called that ship, don't you? The "Starvation Ship!"

We went from San Francisco to Hawaii, and from there straight on down to Wellington, New Zealand. When we crossed the equator from north to south, those of us on the ship who hadn't crossed the equator before passed from pollywogs to shellbacks. We waited below decks for a summons to appear before King Neptune. I was just twenty years old and I had just been promoted to sergeant. At that time, a guy waited a long time to get any rank at all. Guys would spend years just waiting to be promoted to PFC. Here I was just two years in and already a sergeant. So anyway, when I got my summons to appear before King Neptune, it read: "To the guy impersonating a sergeant." I still have my summons in one of my dresser drawers.

Chapter 4

The Birth of Malaria Control

Noumea, New Caledonia

EARLY IN 1942, virtually all of the islands in the South Pacific had long been colonized and were firmly under the control of one of the European Powers. Just as they had divided the continents of Africa, North and South America, and China, so too had the European powers extended their reach across the Pacific. Great Britain, Holland, France, Germany, Japan, and the United States all had interests and possessions from one end of the Pacific to the other.

Our discussion for the moment, though, deals strictly with the French, who were well established in the South Pacific with island possessions spread out from one end of Polynesia to the other. One such colony, New Caledonia, played an important role for the Allies as the war effort grew. New Caledonia was once a penal colony for the miscreants of France and later exploited for its large deposits of nickel. At one time, the nickel mines on New Caledonia produced 14% of the entire world's nickel supply.

French colonists, having moved in over the years, were enjoying the fruits of their orderly colonization in 1942. They operated the nickel mines and managed plantations, grew coconuts and cocoa, or raised cattle on huge spreads reminiscent of the Old West. Imagine the astonished American servicemen seeing native cowboys with ebony skin and frizzy, dyed-red hair standing straight up on top of their heads, riding the New Caledonian ranges and herding cattle like their Texas counterparts halfway around the world.

Only 250 miles long, New Caledonia was an island paradise offering magnificent beauty with broad coral beaches, wooded plains, sharp jagged mountains, and innumerable streams teeming with fish. Unlike many of the neighboring islands in the region, New Caledonia provided a pleasant environment with a friendly indigenous population, relatively free of the multitude of strange diseases that plagued most of the nearby islands in the region.

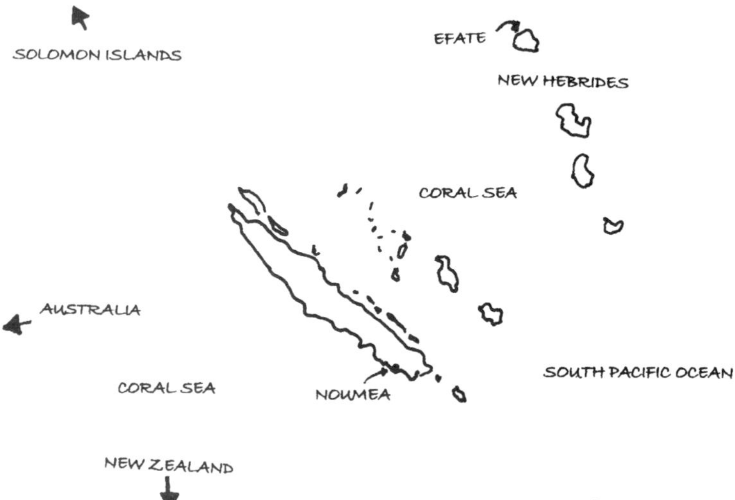

NEW CALEDONIA

For years, however, the locals were looked upon by the French as poor workers, unfit to staff their plantations and too lazy to put in a good day's work. French colonials, therefore, imported laborers from their other possessions throughout the Pacific region. Plantation owners even went so far as to set up a system of indenture similar to that used by the British in colonizing America. Laborers were bound by contract to work a prescribed number of years. In return, they received room, board, and minimal wages for the term of their servitude. An indentured laborer was treated as a slave and could only be freed from his obligations by serving the full term of the indenture.

In 1942, laborers from the French colony of Indochina (present day Vietnam) were deemed the best workers to staff a plantation in New Caledonia. Recruited primarily from the Hanoi and Tonkin Gulf regions, the laborers were nicknamed "Tonkinese" for the area from which they came. Vietnamese laborers soon represented a large percentage of the New Caledonian population. Hard working, industrious, and shrewd in the ways of business, the Tonkinese indentured laborers and servants who made the trip to New Caledonia could work out a ten year contract on a coconut plantation or ranch and upon completion, start fresh with a tidy nest egg set aside from the wages they earned. Many Tonkinese were happy to leave their homeland when they were offered brighter prospects for the future on a tropical paradise in the South Pacific. Little did many of them know how quickly their prospects would brighten even more as hundreds of thousands of U.S. troops and support personnel flooded the islands over the next three years with dollars in their pockets and nowhere to spend them.

Robert Michel, PhM3c USN Malaria Control Unit Cactus. When we arrived in Noumea, our group and the Marines on board were to unload the ship. Our whole outfit, MOB5, stood on the dock while an officer asked us if anybody knew how to do this or that. When he asked who knew how to run a crane, I figured that would be real easy work and I said, "I do." I told him I had run an overhead crane at my uncle's factory where they made roofing shingles. Actually, it was just a hand crane in a roofers' supply building. Anyway, I convinced him that I knew all about cranes, so he assigned me to be in charge of the cranes and told me to get a crew together to operate them.

Having never seen a real crane before in my life, I went back aboard the ship to take a look at them. They were fairly simple. At each hold there were two steam engines and two booms, one positioned over the hold, the other over the dock. Each engine had a wire cable attached to it that went to the boom. By operating the engines, the cable would be brought in or let out. The engines had a valve to turn on the steam, a large lever to raise and lower the cable with the pulley on the end, and a brake. There were two cranes at each of the two holds, one crane for each boom. I tried out the cranes for myself to see how they worked and then picked the crews to operate them. I needed two men per hold, one man on each crane for each shift. I showed them how to do it and away we went, all day and night, unloading our supplies from the ship. The guys down in the hold loading the cargo nets were passing out from the heat and I was walking around "supervising" my crew and drinking coffee. It was the best bluff of my life when I told them I knew how to run a crane. Once in awhile, if you are careful, it's OK to volunteer.

I don't remember too much about Noumea other than unloading the ship, the beginning of the construction of the hospital, and a few liberties we took while there. There was a place up on a hill in town called the "Pink House," a house of "ill repute" as they used to say. I know a lot of guys would go there because the locals in town, who spoke French, were not too friendly.

When United States Navy ships pulled into Noumea that spring and began to unload their troops and support personnel, French authorities and the New Caledonian populace really began to worry. Venereal disease on the island had until now been kept in check by the French Health Service, who imposed stringent regulations for testing and diagnosing the maladies. The introduction of foreign armed forces, however, posed a problem that could easily get out of hand.

General Patch, the U.S. Army commander for the area, responded to local fears by calling a closed-door meeting in March, 1942. The Island Provost Marshal was there, as well as Dr. Arthur King, MD, the U.S. Army Island Medical Inspector. Briefly, Patch outlined a project to establish in Noumea a military bordello that would be constructed for the exclusive use of American troops. To this day it is not known for certain who initiated the project, local entrepreneurs or the U.S. Army. Regardless, Patch described the project as solely a French idea and ordered Dr. King and the Provost Marshal to get all the pertinent details from the French Chief of Police and to cooperate. The general's parting words as related by Dr. King were, "Remember, gentlemen, that no matter what happens I am never to know that the house exists or that its activities are connected in any way with the United States forces."

The "original" Pink House was an old private residence located on the rue Paul Bert. It was purchased by a consortium of French locals that included the Chief of Police and the Intendant-General. The structure was quickly painted pink, hence its infamous name. Modifications were made to the residence, equipping it for the brisk trade which was soon to follow. During the next four months of its stringently monitored operation, it has been said that the domicile accommodated over 24,000 man-visits with not a single case of venereal disease ever traced back to the Pink House.

Arthur King, MD, US Army Surgeon, Medical Supply Officer-South Pacific. It seems that there were actually three Pink Houses in Noumea, one run by the Navy at the docks, the "original" on the rue Paul Bert, and a third further uptown for officers only. I guess you could say it was a case of a recognizable generic name for a useful medical procedure. But no matter what anyone tells you, they existed at different times and were eventually closed down for different reasons.

William Ricciardi, MM2/c Boat Gang USS Crescent City. I remember the "Pink House"—everybody knew of that place. It was on the edge of town and had a real reputation. I never went there myself though. I remember a lot of the guys getting their hands on that "Butterfly Brandy," too.

Robert Michel, PhM3/c USN Malaria Control Unit Cactus. The local bars on Noumea would sell the troops "Butterfly Brandy." It was the worst tasting concoction ever made by man. It had the exact taste of kerosene and it was rumored around the camp that a couple of guys went blind drinking it. Who knows?

Dave Minton, PhM3/c, USN Malaria Control Unit Cactus. Did Bob tell you about the "Pink House"? Boy, that place was something else. We'd go by there and see the guys just lined up waiting to get in. Can you imagine all those guys going ahead of you like that? Thank God I never got in a situation where I ever had to pay for it.

The Growth of Malaria Control—from Roses to Cactus

What was probably the easiest, and the most enjoyable aspect of writing this book, was the time I spent with Bob Michel and the other guys—listening in awe to their stories and transposing them to paper. What proved to be most elusive and frustrating, however, was fleshing out the story of malaria control itself. Day after day, week after week, I would find myself proofing the rough draft, adding new stories, talking with more veterans about their personal experiences; anything in an effort to convince myself I was making progress in finally bringing the project to completion.

As weeks turned into months, with no real definition to the story taking shape, my frustration finally peaked as I found my malaria control research repeatedly turned up either conflicting narratives or complete dead ends. On the one hand, I had located what seemed to be a credible Navy Medical Corps History; credible that is, until compared with the official Army Medical Corps History. Taking into account Bob Michel's stories, along with those of Dave Minton, Dr. Arthur King, Bill Schultz, and others, I felt I could fill a 55 gallon drum with the paper my notes were written on and still not come up with a single page of coherent text to add anything meaningful to their story. Many a sleepless night was spent churning through every piece of information I had, and probably an equal number of nights were spent writing imaginary lists of questions which I was sure, once answered, would bring closure to my mission.

Late one night, I steeled myself for a final look at everything I had. One way or the other, I had to put this to rest. I would reach my arms out, only to grab nothing more than a cloud of smoke. So, this was going to be it. I laid out a time line starting in 1941 and stretched it out all the way to 1945. Using verified events, with known dates, I started the slow process of putting each piece of information, from every source I had, into chronological order. A little over four hours later, my chart was complete. Like the sun popping through the clouds after a hard spring rain, the story of malaria control lay right before my eyes. Perfect in every detail? Probably not. Accurate? Probably 80%, at best. What I did end up with, though, was a pretty fair idea of how and why Malaria Control Unit Cactus came to be, when and where they went, and the contribution, albeit unknown at the time, that its members made in their own small way to the war effort. I came away from the exercise with a feeling I had somehow validated not only the existence of the unit itself, but also the individual efforts of hundreds of guys like Bob and Dave who performed one of the lowliest, dirtiest, least recognized jobs in the entire war.

The history of malaria control in the South Pacific proved difficult to document for a number of reasons. First, the area was divided into two regions, with control of those regions divided between two branches of the military, thus resulting in two disparate histories. The Army, under the command of General Douglas MacArthur, was given responsibility for Australia and all the islands west and north of the Solomons, including New Guinea. The Navy, commanded by Admiral Ghormley, took control of everything else east and north of Australia and the Solomons. This gave the Navy, with all seagoing assets, responsibility for the French protectorates of New Caledonia, the New Hebrides, Fiji, Samoa, and the critical Solomon Island archipelago. When troops and supplies started flooding into the area to build supply bases and airfields, overall command of all branches of the military and their activities in the area fell to the Navy. The Eastern South Pacific, essentially, became a Navy show from the very beginning of 1942.

Recognizing the impact tropical disease would have on the exposed troops, the Navy assigned a high ranking officer to outline a strategy for the survey, eradication, and control of vector borne disease (insect transmitted illnesses such as malaria and dengue fever). To make sure he had the authority needed to carry out his mission, he was placed on the staff of Admiral Ghormley, Commander South Pacific. Well connected all the way back to Washington, the high ranking Navy officer was a career medical doctor by the name of Lieutenant Commander James Sapero. Given absolute authority over planning and executing a strategy, Sapero early on recognized the need to isolate the troops from the heavily infected natives on all the occupied islands. He also needed trained medical personnel to survey and identify the population concentrations and areas most infested with the known malaria-carrying *Anopheles* and the dengue-infected *Aedes aegypti* mosquitos. Finally, he recognized the need to organize and train teams of laborers to eradicate suspected insect breeding grounds and to control the spread of disease with the limited supplies of quinine and synthetic drugs at his disposal.

As the campaign in the South Pacific expanded, so too did Sapero's mission. Not only was he concerned with those islands known to be infested, the New Hebrides in particular, he now had the added responsibility of keeping the non-malarial islands—New Caledonia, Fiji, New Zealand, and Samoa—free of infestation. As the Allied offensive progressed, traffic to and from the islands dramatically increased the risk of spreading malaria and dengue fever from the infested islands in the combat zone to those islands heretofore free of these mosquito-borne diseases.

When Admiral Ghormley transferred his flag to the shores of Noumea, New Caledonia, on June 15, 1942, he also started building an infrastructure to support his ever growing fleet and their needs. An integral part of that infrastructure was the construction of medical facilities and the drafting of personnel needed to care for his naval personnel. By June, 1942, the Army, having started their landings on New Caledonia in April, had already established three fully operational hospitals in the immediate vicinity of Noumea. What would later turn out to be a key factor in the progress and outcome of malaria control efforts in the South Pacific was the steady increase of the U.S. Army presence in Sapero's area of responsibility. It didn't take but a few months of steady reinforcement for the Army to far outnumber the Navy, causing an enduring friction between the two branches. Sapero issued the orders and expected the Army to carry them out implicitly.

To the north on the New Hebrides, Marines and Navy SeaBees had landed as early as March, 1942, to build an advanced airfield to defend the islands. The malaria rate on the island of Efate, which is part of the New Hebrides, at that time rose beyond epidemic proportions. Marines and SeaBees started dropping in astonishing numbers. It was estimated that the rate of malarial infection on Efate that spring rose to a level of 2,600 cases per 1,000 men per year. To translate that statistic, virtually every man on the island was infected 2.6 times per year. Clearly these numbers represented casualty figures far in excess of acceptable, as hundreds of troops and construction workers were lost to the war effort without ever meeting the Japanese enemy in battle.

By the time Bob and Dave arrived in Noumea, New Caledonia, around June 15 with MOB5, Lieutenant Commander James Sapero was just starting to set up shop at a Naval installation adjacent to a nearby beach on the edge of the city. His first order of business was to organize the survey, control, and eradication teams needed to quell the epidemic raging on the New Hebrides. Looking ahead, his mission would soon extend to the combat zone in the Solomons where Allied forces were scheduled to launch their invasion as early as August. Events were rapidly outstripping planning efforts, leaving Sapero little more than six weeks to organize and deploy his personnel, equipment, and supplies.

His strategy encompassed a three-phase approach to the problem. In the first phase, survey teams were to identify the diseases the troops faced, the level of infection in natives near or adjacent to Allied troops, and the rate and type of infection plaguing combat and support units. The second phase of his strategy called for the formation, training, and deployment of control and eradication teams to interrupt the vector of insect transmitted diseases by destroying their breeding grounds and habitat. The third and final phase of his strategic operations would include the treatment and prevention of vector borne diseases in the rank and file of Allied forces in the islands. Prevention would be achieved by mandated, liberal doses of the malarial prophylactic Atabri F e and the use of limited supplies of quinine to treat those most severely stricken.

Since early 1941, anticipating hostilities in the Pacific, the Navy had been training entomologists, malariologists, and medical doctors at Bethesda Naval Hospital in Maryland to develop specific expertise in identification and treatment of vector borne diseases. The Navy Medical Department even went so far as to establish special units for these specialists which they designated the Hospital Volunteer Special Group, H-V(S). In May, 1941, the first Navy entomologists were commissioned in the program—Ltjg. William Lawlor and Lt. Paul Woke.

It was personnel from the ranks of the H-V(S) units that Sapero would rely on, less than a year later, to populate his original Malaria Survey Teams. The responsibility of the Navy Malaria Survey Teams would be to move into combat zones and areas of occupation and very rapidly locate and identify any vector borne diseases prevalent in the native populations and in those areas adjacent to military bases. Trained doctors and medical personnel would take blood samples from the locals to determine levels of infection, while entomologists would look for the *Anophleles* and *Aedes aegypti* mosquitos to identify their likely breeding grounds. In addition, Sapero's survey teams would ultimately be responsible for identifying and isolating many of the other killers running rampant in the region which included elephantiasis, leprosy, tetanus, meningitis, and hog cholera. Survey teams ultimately saved thousands from tropical infection by identifying areas safe for bivouacs, quarantining particularly dangerous areas such as the tetanus infected burial grounds on the New Hebrides, and isolating native populations who posed particular risk from contact with Allied troops.

When Lt. Cmdr. James Sapero arrived in Noumea in the spring of 1942, entrusted with a letter of authority from the highest levels of ComSoPac, he was accompanied by a trained Navy entomologist, Ensign Kenneth Knight, and a medical doctor, Ensign Robert Mount. From this original group of three men, the ranks of Navy Malaria Control personnel would swell to almost 9,000 medical officers and enlisted men by August of 1945 when the Japanese were finally pummeled into submission. All three men would eventually survive the war, serving together in the

Navy as late as 1958, when they were again dispatched by the U.S. Bureau of Medicine to Cairo, Egypt, to battle one of the world's worst epidemics of malarial infection.

Organizing the first survey team at Noumea under the direct guidance of Ensigns Knight and Mount, it only remained for Sapero to fill the ranks of his eradication and control teams with laborers. Given the military's overall shortage of personnel in the South Pacific, it was left to Sapero, Knight, and Mount to seek warm bodies from the newly arrived medical units such as MOB5. To effect their transfers to his planned malaria control teams, Sapero would have to find men willing to volunteer. To this end, therefore, LtCmdr. Sapero painted a glamorous picture of an elite, top secret outfit he was forming that would be heading "up the islands," straight into the teeth of the action. When presented to naïve, young men like Bob and Dave who had spent the better part of the last year and a half in the boredom of peacetime military life, Sapero's offer was everything they wanted to hear—the opportunity for glory and the chance to meet the hated Japs. It didn't take Bob and Dave but a moment to jump into Sapero's outstretched arms to join the ranks of his malaria control and eradication teams.

Under the direction of his survey officers, Knight and Mount, Sapero's control teams would soon lead the point of attack for phase two of his offensive. Backpack sprayers loaded with No.2 diesel oil would soon become Sapero's weapon of choice for control and eradication personnel. They were trained to spread a thin film of oil on any suspect body of water that might harbor the Anopheles' larvae. Axes and shovels (later bulldozers and front end loaders) hacked away any jungle growth suspected of harboring mosquitos, thereby denying the *Anopheles* safe haven.

With the power that came from his position on Admiral Ghormley's staff and the authority proffered from the highest offices of the Navy, Sapero issued the orders necessary to launch the third phase of his plan—treatment and prevention of disease.

Atabrine, the quinine synthetic, lay at the core of his prevention program. All troops stationed in any suspect area were to be treated weekly with Atabrine to minimize the possibility of infection or, at the very least, mask the symptoms once infected. Quinine, available in limited quantity because the Japanese had swallowed up most of the quinine producing countries in the Pacific, would be administered only to those most severely disabled by malaria.

By early August, the first of Sapero's survey and control teams were dispatched from Noumea arriving at Efate in the New Hebrides on August 12, 1942. Under the command of Ensign Kenneth Knight, the group included Ensign Robert Mount, Pharmacist Mates Robert Michel and Dave Minton, and an odd assortment of volunteers. Just one day later the original Malaria Control Unit was divided and two units were formed: Malaria Control Unit Roses with responsibility for malaria control activity on the New Hebrides and Malaria Control Unit Cactus led by Ensigns Knight and Mount slated to leave that morning for detached duty with the 1st Marine Division on Guadalcanal.

Malaria control unit Roses immediately launched malaria survey and control operations from a Quonset hut near the airfield at Vila on the island of Efate. Ensigns Knight and Mount, with their team, which included Bob and Dave, loaded up on an Army transport and headed "up the islands to where the action was," just as Sapero had promised. They landed at Henderson Field on Guadalcanal just days after it was opened to Allied air operations.

Robert Michel PhM3/c, USN Malaria Control Unit Cactus. Not more than two weeks or so after we arrived at Noumea, my buddy, Dave Minton, and I were down at the beach for a swim. The beach was actually part of the hospital compound. We didn't have to go on liberty to go there. Dave and I ended up talking to a Doctor by the name of Lt. Cmdr. James Sapero about a group he called Malaria Control that he was forming and that was heading up into the islands. He was looking for volunteers. Sapero really didn't tell us much about what we were going to do, it was more like, "How would you guys like to go up in the islands where the action is?" So, being young and stupid, Dave and I volunteered. When we told him we would go he simply said, "OK." The next thing we knew we were headed for Efate in the New Hebrides.

Arthur King, MD, US Army Surgeon, Medical Supply Officer-South Pacific. I remember Sapero. He was a man who knew everything and nothing, all at the same time. He was going to wipe out malaria from the islands down there all by himself. He really knew nothing about malaria. I stayed as far away from him as possible.

I remember Malaria Control Unit Cactus. There were eight men in the original unit. I stayed away from them too. They didn't know what they were doing, but how could they? They were given no training and little in the way of equipment. Sapero was a guy that believed in heavy equipment: sprayers, shovels, axes, and the like. That's what he equipped those guys with. In fact, I don't think he ever even went over there to Guadalcanal to see what those guys were doing.

Efate: Starting Up the Islands

Referencing geography of 1942, Efate was one of over ninety islands that made up the New Hebrides Condominium. Located in the center of the north south stretch of the New Hebrides Islands, Efate was blessed with a huge deep water harbor, giving it significant strategic value to the precious shipping lanes between Australia, New Zealand and the United States.

Originally charted by the Spanish in 1605, and in later years traveled by the French and the English, subjects from both countries were quick to settle on some of the most beautiful islands in the South Pacific. It wasn't until 1906, however, that both England

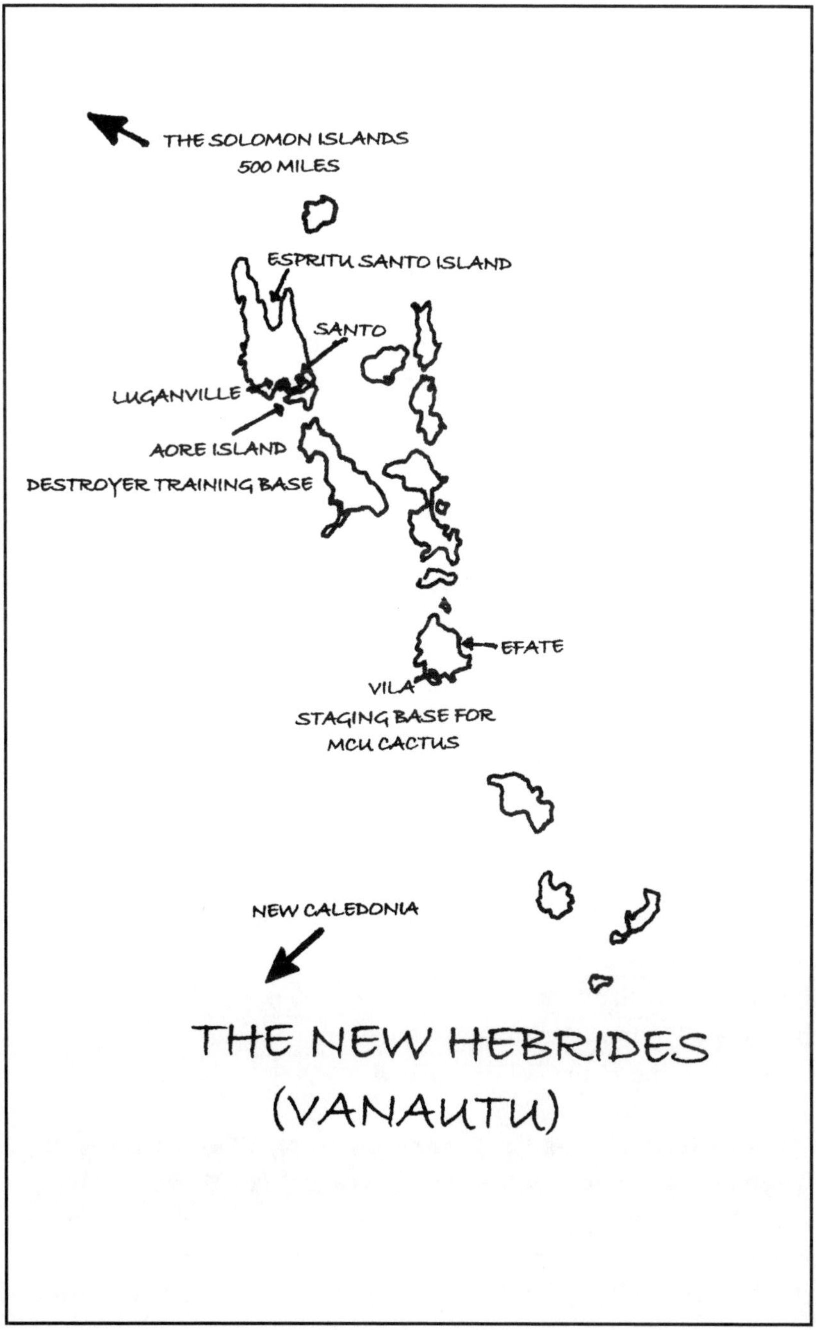
THE SOLOMON ISLANDS
500 MILES
ESPRITU SANTO ISLAND
SANTO
LUGANVILLE
AORE ISLAND
DESTROYER TRAINING BASE
EFATE
VILA
STAGING BASE FOR
MCU CACTUS
NEW CALEDONIA
THE NEW HEBRIDES
(VANAUTU)

and France signed an agreement settling their claims on the archipelago and making it the only joint ruled Condominium in the entire world. This arrangement lasted through World War II and up until 1979 when free elections were finally held and the independent nation of Vanuatu was declared. Looking for the New Hebrides on a current map of the Pacific will only frustrate the reader, one must today look for Vanuatu.

Located just 1,200 miles northeast of Australia, the New Hebrides, along with New Caledonia to the immediate south, were essential to U.S. and Japanese interests in the area. As the Japanese spread farther east and south from New Guinea moving into the Solomon Islands in May, 1942, the ports of Noumea and Espiritu Santo in New Caledonia had to be held and reinforced at all costs by the Allies to protect Australia and New Zealand. U.S. forces flooded the area, building bases and port facilities on many of the islands. Efate, Espiritu Santo, Noumea, and Wellington, New Zealand, became vital staging areas for the intended Allies' move to block Japanese expansion in the South Pacific.

William Ricciardi, MM2/c Boat Gang USS Crescent City. In February, 1942, we loaded up 500 Marines and five 5" guns at Pearl and headed down to Efate in the New Hebrides. That was way back in February; we didn't hit the Canal until August. In fact, we charted that harbor. When we got there, there weren't even any charts of Efate harbor. From there we went to Suva in the Fijis and with another ship unloaded 200 mines. The only ship that beat us in there to Suva was the USS Chicago. We were the second ship there, that's how early it was.

John Richter, PhM3/c Corpsman USN, 75mm Gun Half-Track Headquarters & Service Battery, 11th Marines. Many of we Navy Corpsmen and Marines had, just prior to our assault on the Canal, sailed on the "Starvation Ship," the SS Ericcson. This civilian vessel had failed miserably in an attempt to take on a sufficient supply of food for our trip from California to New Zealand. It is officially documented that many of us lost from 16 to 23 pounds during the trip.

Marine Sergeant, Section Leader 75mm Pack Howitzers, I/3/11. While we were in Wellington, we transferred everything off the Ericcson and loaded it onto the McCawley. It was right at this time that the stevedores in Wellington went on strike, so we had to do all the loading ourselves.

Dave Minton PhM3/c, USN Malaria Control Unit Cactus. I remember that once in awhile when we were at Efate, a few Jap planes would come over and bomb us. Not many, and not very often, but they would get through once in a while.

One such base, located at Efate, New Hebrides (code named Roses) was United States Naval Hospital No. 2. Built above the harbor on a hill overlooking the town of Port Vila, Roses was to play an important role in treating U.S. wounded evacuated from the ground, sea, and air combat taking place further up the island chain on Guadalcanal, Tulagi, and the Floridas. Unknown to but a few in the military command, Roses was also the staging area for the newly formed Malaria Control Unit Cactus. It was here in a non-descript Quonset hut on the grounds of Naval Hospital Roses that members of MCU Cactus were outfitted for their impending deployment to Guadalcanal in the Solomon Islands.

Robert Michel, PhM3/c USN Malaria Control Unit Cactus. Shortly thereafter we were transferred from MOB5 to Malaria Control and were flown from Noumea to Efate in the New Hebrides. There were more volunteers there already, not just Dave and I. I always assumed these other guys arrived the same way we did, from some other installation somewhere.

We were only there for a short time, a day or two, when we were taken to a big Quonset hut where we were outfitted with weapons and other equipment. The 03 (Springfield Model 1903 30/06 bolt action rifle) and the .45 (Colt .45 caliber Automatic pistol) with a holster and a web belt. We were also issued a canteen, our mess gear, and all the rest of the stuff. Oh yes, we were each given a pith helmet. The reason medical personnel were issued weapons was that the Japanese were considered a barbaric enemy, so even though we were technically Corpsmen, we were allowed weapons. That's about all we were outfitted with. The doctors, one entomologist and one MD, were also equipped with microscopes and slides.

As far as the microscopes were concerned, I didn't know what they were going to use them for unless it was some kind of research. It turned out later that we used them to look at slides of blood taken from infected people to identify the infecting parasite. I remember there were two or three kinds of malaria, and one could be deadly.

Malarial infection in humans actually occurs in four different forms, depending on the species of the parasite. Each of the four types of malaria have the same symptoms including sweating, fever, and chills. Malaria left untreated will reoccur periodically over long periods of time. MCU Cactus was most concerned with the two types of malaria indigenous to Guadalcanal and, ironically enough, one of these included the deadliest strain that had potentially debilitating effects.

The first, and probably the mildest form of malaria, is that caused by the *Plasmodium vivax* parasite. It is often referred to as benign tertian malaria. The initial attack of fever usually occurs about two weeks after infection and may reoccur every two days, sometimes even returning years after the initial mosquito bite.

Jungle fever—or malignant tertian malaria—is the second and most dangerous form of malaria endemic to the Solomon Islands. It is caused by the *Plasmodium falciparum* parasite. Often blocking the blood vessels leading to the brain, *Plasmodium falciparum* frequently produces coma, delirium, and in cases left untreated, death. It was the job of MCU Cactus to continually check blood samples of infected Marines on Guadalcanal in an effort to identify the diseases present and try to thwart the most serious effects of the parasites. Jungle fever was the cause of most of the deaths that occurred as a result of malarial infection, both for U.S. and Japanese troops on the island.

Robert Michel, PhM3/c USN Malaria Control Unit Cactus. Our group was made up of two doctors and six or eight enlisted men who were Corpsmen and Pharmacists Mates. We had a Chief by the name of Rountree who kept us all together. We were all volunteers with absolutely no training or preparation for the job we were to do. The Malaria Control Unit was formed, I believe, as an afterthought on the part of the Navy.

When we were set to go, my god, I was excited. We were about to set off on the great adventure. We couldn't wait to get where we were going even though I don't remember ever being told the name of the island we were heading to. We just knew we were finally heading to where the action was.

Marine Sergeant, Section Leader 75mm Pack Howitzers I/3/11. I was aboard the USS McCawley with my unit, I Company, 3rd Battalion, 11th Marine Regiment, when we attacked Guadalcanal; that was August 7, 1942. I was just twenty years old at the time and just a kid. We were all kids. People have this impression of the Marines that we were all hard drinking, women chasing, tough guys. That is totally wrong. We were a bunch of kids. I was what you would call a "pogey bait" Marine, a kid who loved to get candy and gum at the PX. I never went out drinking or chasing women; I didn't know the first thing about it. In fact, most of us hadn't even kissed a woman other than our sisters, mothers, or grandmothers when we went in there. Most of us had never even known a woman. I was still a virgin when that whole thing started.

I was a Section Leader in charge of three 75mm Pack Howitzers. Our Task Force approached the island from the southwest and headed toward Savo Island. The ships then split into two groups, one group headed for Tulagi and Gavutu and our group veered off toward Red Beach on the north coast of Guadalcanal.

The Marines that went into Tulagi that morning ran into a lot of resistance and had a real tough time of it; we didn't. There was only a construction crew and a few Koreans working on the airfield that took off when we landed. I was in the third wave that went

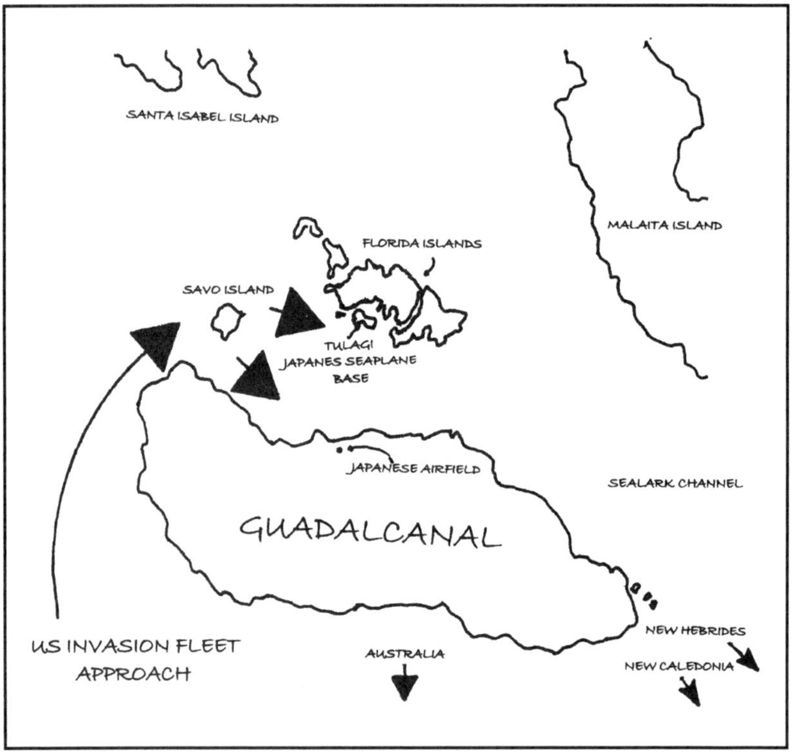

in; that's the third wave, you know. The only casualty our group took on the first day was a guy who was cutting open a coconut with a machete and cut his thumb off. Can you believe that?

Things changed pretty quick after we got there though.

Lt. Cmdr. Lewis P. May, USNR Ret., Assistant Supply Officer, USS President Hayes APA-20. It was calm where we anchored at Tulagi that morning. Our unloading was halted, though, when we heard fierce fighting ashore. During the night, the situation was obscure but serious. Then we got the news that enemy ships were on the way and although we still had the major portion of supplies and provisions for the Marines still on board, we were ordered to pull out unescorted. So we left at top speed. I think it was almost a month before we were able to return with supplies and food by which time the Marines were eating Japanese rice and K rations."

William Ricciardi, MM2/c Boat Gang USS Crescent City. We went in there in the first hit on the Canal in August. The problem was, we were given at most 15 hours to unload. The ships had to be out of there before the Tokyo Express pulled in at night. If you weren't done, the ships pulled out without you. I was left on the beach many times. Once I was trying to get back to the ship and I was so close, I kept telling them to throw me a line. The officer on deck told me to sheer off. I called for a line again, and he yelled back if I didn't sheer off they would run me over.

Marine Sergeant, Section Leader 75mm Pack Howitzers, I/3/11. I don't think most people know how much of a shoestring operation that whole thing was. I mean, we really didn't have any of the good equipment or anything. We were issued the old '03 rifle, if you remember, the old long bayonets, and even the yellow grenades with the seven second fuse left over from World War I. When we used those old grenades, the Japs could see those at night and with a seven second fuse, they would just pick them

NAVY DEPARTMENT COMMUNIQUÉ

August 10, 1942
South Pacific

A STATEMENT BY ADMIRAL ERNEST J. KING, COMMANDER IN CHIEF, U.S.FLEET

1. Offensive operations by U.S. naval and other forces, looking to the occupation of islands in the Tulagi area in the southeasterly Solomon Islands, have now been underway for about 3 days.

2. The operations are under the immediate command of Vice Admiral Ghormley and under the general control of Admiral Nimitz. Certain of the forces under General MacArthur are cooperating.

3. The objective of the current operations is to expel the Japanese from the Tulagi area and to make use of that area for our own purposes. The enemy have been in process of consolidating their positions, in which their purpose has been not only to deny them to us but to use them as a base of offensive operations against our positions which cover the line of communications to Australia and New Zealand.

4. An initial surprise was effected and planned landings accomplished. The enemy has counterattacked with rapidity and vigor. Heavy fighting is still in progress. Our operating forces are employing all available communications in the conduct of the operations, so that our information is incomplete but it appears that we have had at least one cruiser sunk and two cruisers, two destroyers and one transport damaged. Likewise, information as to the extent of damage to the enemy is incomplete but includes a large number of

> enemy planes that have been destroyed and surface units put out of action.
>
> 5. This operation in the Tulagi area is significant in that it marks our first assumption of the initiative and of the offensive. Any of the previous operations in the Pacific, however successful, have been essentially defensive in character.
>
> 6. It should be understood that the operation now underway is one of the most complicated and difficult in warfare. Considerable losses such as are inherent in any offensive operation, must be expected as the price to be paid for the hard-won experience which is essential to the attainment of far reaching results.

up and throw them right back at us. It wasn't until much later that we started getting the newer green grenades with the three second fuses, the short bayonets, and the M1 rifle.

Everyone wrote us off two days after we landed on the island. When we hung on, we surprised everyone, the brass, the politicians, and the press.

John Richter, PhM3/c Corpsman USN, 75mm Gun Half-Track, Headquarters & Service Battery, 11th Marines. With little resistance from the enemy, Guadalcanal seemed at first to me like a "travel-brochured" South Pacific Island, with additional, FREE, fresh coconuts available for the taking! Man, how in time that peaceful setting reversed itself.

Robert Michel, PhM3/c Malaria Control Unit Cactus. To the best of my recollection, we were then flown in to the Canal on the 13th of August, 1942. My recollection is that it was broad daylight when we landed. The plane we were on was what we called, at the time, a DC3. Later the military designated these planes either the C47 or the R4D depending on the unit. The interior of the plane had a long aluminum seat with indentations for sitting along each side of the plane facing the center. I remember all of us sitting in those bucket seats on the airplane looking at each other, heading for who knows where.

As I recall, the flight from Efate to Guadalcanal's Henderson Field was an eventful one, to say the least. The longer we were in the air, the more apparent it became that the pilot was lost. Navigation over the Pacific Ocean in 1942 was not what it is today. We weren't even sure we were going to have enough gas to find the island and land.

While this was going on, an electrical fire also started in the plane. That only added to our stress as we all chipped in to find the source of the problem and get the fire out. As it was, there was no significant damage to the plane but we thought for sure we weren't going to make it.

The first U.S. aircraft that landed at the newly constructed Henderson Field arrived on August 12, 1942. That day, the airfield was inspected by Admiral McCain's aide, Lieutenant Simpson, who declared the field was ready for flight operations. Ironically, it was Admiral McCain's personal Catalina that made the first landing on Guadalcanal.

Dave Minton, Phm3/c, USN Malaria Control Unit Cactus. I don't recall much about the landing other than the airfield seemed deserted. There was no one around. All we saw when we came in were the wrecks of a few destroyed planes scattered around the runway.

William C. Dickman, MSgt. USMC Enlisted Navigator/Naval Aviation Observer VMJ-253, South Pacific Combat Air Transport, SCAT. We departed for the South Pacific on August

NAVY DEPARTMENT COMMUNIQUÉ

No. 104, August 12, 1942
South Pacific

1. Operations in the Solomons are still in progress.

2. It has been confirmed that the United States Marines have landed as scheduled on three islands in the vicinity of Tulagi. The Marines are engaged in consolidating their positions. Supporting naval forces have been engaged in bitter fighting, details of which are not yet available. United States Army and Australian aircraft are continuing attacks on the enemy landing fields and shore based aircraft.

3. There is substantial evidence that the Japanese had planned and had well underway the development of an enemy base in the Tulagi area.

4. Until more details are available it is impossible to elaborate further on these operations. All available communications facilities are overtaxed by urgent messages concerning operations.

23, 1942, and our entire unit made a mass flight from North Island in San Diego, California, stopping at Hawaii, then on to Samoa, the Fiji Islands, and finally New Caledonia. We were the first transport squadron to arrive in New Caledonia. We immediately started operations the 1st of September, 1942. My squadron was stationed at Tantouta, New Caledonia, and I was assigned to VMJ-253, MAG-25. My position on the air crew was that of Enlisted Navigator or Naval Aviation Observer.

Chapter 5

Memories of the Canal

And Other Interesting Places We Visited

Imagine for a moment that tomorrow you are asked to go to a place that you have never heard of before. A place for which no maps exist and little is known of the weather, the geography, or the people. For all you know the place is inhabited by a tribe of headhunting cannibals. You will be completely isolated from your family, friends, all news, radio, television, and your computer. When you leave you don't even know how long you will be gone and everything you own and need for survival is in a pack on your back. Add to this the one thing you do know, the place you are going is occupied by an unknown number of people who not only hate you but are hell bent on killing you when you get there. It wouldn't take much thought to say "No" to that, now would it?

The actions that took place in the Solomon Islands from early August 1942 through February 1943 encompassed thousands of square miles of territory, hundreds of Pacific islands unknown to most, and every branch of the United States military. Tens of thousands of combatants on both sides of the conflict participated; Japanese, Koreans, Americans, Australians, British, New Zealanders, Fijians, Tongans, Vietnamese (Tonkinese), Solomon Islanders, Chinese, Dutch, French, and Indians. Well over a hundred books have been written on the subject, presenting the strategic and tactical levels of the campaign in the detail it deserves as the pivotal campaign in the Pacific War. Although Bob's story and that of the Skeeter Beaters is a great one, the reader is encouraged to explore the campaign in greater detail with the literature and their living forebears to gain a broader understanding of the truly extraordinary times these men lived in.

The story of the Skeeter Beaters is, at best, a microcosm of much larger events that were being played out during this time. What Bob, Dave, and the others portray is not only their "take" on what they saw and were asked to do, but more importantly what they can still remember 58 years later. Most participants in the Solomons campaign were never afforded the luxury of a broad view; most of them never even knew where they were going much less what they were to do when they got there. These were just ordinary men who were asked by their peers to perform extraordinary feats amidst the reality of life-changing events. Dozens of veterans who were contacted for comment on this story very clearly stated to the effect, "I was just a private. I was given a job to do and I did it. I never knew what was going on. I always assumed our officers knew that." What is important to this discussion, however, is the fact that though very little was known about this area of the Pacific, at the time, these men unquestioningly went where they were told to go and performed their duty as dictated.

One fact that was known by U.S. military planners at the time, however, was that the Japanese were constructing an airfield on the northern plains of Guadalcanal, which if completed, would alter the entire military balance in the region. Vital lines of communication and supply between the United States and Australia lay directly in the path of enemy bombers that were slated to be based at that airstrip. Invasion of Australia and New Zealand would be the next logical step for the Japanese to take in their attempt to dominate the Pacific.

The islands that Bob, Dave, and twenty thousand other young men stepped onto that August were a veritable paradise when viewed from the air. At ground level, however, they would soon experience the harsh reality of close quarters combat in one of the world's wettest locales, one which averaged as much as 164 inches of rainfall each year. They would experience only two seasons, the wet and the dry, and the timing of their arrival was marked by the beginning of the wet season. Add to the inhospitable clime an enemy just as determined to emerge victorious as they were, and the unwitting young Americans were to experience seven months in the worst hell imaginable. I often think it was best that the young Marines, sailors, and aviators of America didn't really know what they were getting themselves into. Had they been given the least glimmer of what their lives would be like in the weeks ahead, perhaps the course of history would have taken a different turn, one with a less than positive outcome.

What a pity that just 58 years after our fathers and grandfathers walked, crawled, or were carried off that stinking hellhole, that many of their own children and grandchildren don't even know what Guadalcanal "is" or where it is located. Few are the descendants who, today, realize the significance that the events that took place on such a small island half way around the world would play in the lifestyle they take for granted and freely enjoy today.

When U.S. Marines landed on Guadalcanal and adjacent Tulagi on August 7, 1942, the move represented the first major offensive by combined American military forces in World War II. Their subsequent engagement with Japanese forces was to become a campaign that lasted for over six months, during which time the outcome swayed precariously back and forth between the opposing forces, the ultimate conclusion daily hanging in the balance.

The Guadalcanal campaign included six major naval battles the likes of which the world had never before seen and will likely never witness again—daily air engagements and continuous skirmishes between ground forces all over the northern coast of the tiny island.

No campaign in either the Pacific or European Theaters of Operations, throughout the course of World War II, was witness to such continuous action over such an extended period of time by combined naval, air, and ground forces.

Marine Sergeant, Section Leader 75mm Pack Howitzers I/3/11. Every man that served on Guadalcanal was in combat every single day he was there and it went on for months. He was in danger of being killed literally every day either from the bombing, the shelling, the snipers, Jap infiltrators, or close combat. It was a very dangerous place.

William Ricciardi, MM2/c Boat Gang USS Crescent City. Every day there was combat either on the beach or in the waters around the Canal. You could tell the Japs had this confidence about them that they would kick us out of there any time. One time I was on the beach and a Jap cruiser just steamed right into the channel and started shelling us. I could see the crew lining the rails and they were fishing off the side of the ship while they were shelling us. Fishing! If that isn't confident, I don't know what is. Jap subs would operate in the channel there and we could see them when they came up. They would surface and put a boat over the side and just come right into shore and then do their business and head right back out to the sub. That went on all the time. The Canal was a crazy place and there was a lot of stuff going on there that you never really hear about.

Dave Minton PhM3/c, USN Malaria Control Unit Cactus. I remember the place really was a shit hole. I remember stink and smells and so many times while I was there I would ask myself, "What the f@#! am I doing here?"

Marine Sergeant, Section Leader 75mm Pack Howitzers. I/3/11 There was always a smell there of death. You could smell it no matter where you were; you could even smell it in the channel on the ships. You know, we didn't bury any of the Japs we killed, we just left them lay where they fell. The place really had a smell that I can remember to this day.

Robert Michel, PhM3/c Malaria Control Unit Cactus. When we got to the Canal, we were not attached to any hospital or medical facility that I know of. We were independent of everyone. Even the 1st Marine Division people left us alone. We simply had no one to answer to except our own officers and they were doctors, not career Navy.

Dave Minton PhM3/c, USN Malaria Control Unit Cactus. No, we really didn't have anyone to answer to and no real military regimen or duties to speak of. We had our job to do which was our duty, you know, and that was about it. No watches or anything like that at all. Just what we were supposed to be doing with Malaria Control.

Malaria Control Unit Cactus, soon to be referred to in the months ahead as the "Skeeter Beaters," had just entered the war.

Robert Michel PhM3/c, USN Malaria Control Unit Cactus. We were just a pickup group of Corpsmen from New Caledonia and who knows where else and just given the name of Malaria Control. We were not assigned to any medical facility but were rather an independent group assigned to the 1st Marine Division, but with no direct command. We had no insignia, patches, or anything else to distinguish us from any other corpsmen on the island. I do recall one of our officers saying that we had a letter from Admiral Nimitz that gave us top priority to fly anywhere in the South Pacific. I don't know if anyone actually used it or not. Usually, if we had to go to another island we just hitchhiked on a plane that was going where we needed to go.

Anyway, when we arrived on Guadalcanal, we were given everything we needed, including a jeep and tents. We set up our tents in an area called Lunga. I remember there was a coral road about 50-100 feet in front of our tents that went toward the Matanikau River. (The Government Track or Coast Road was a crushed coral road that ran east to west along the northern coast of the island just inland from the beach.) If I remember correctly, to get to the hospital, we would go out our tents to the road and make a left turn heading off that way.

We were far enough away from Henderson Field so that Japanese shells and most of their bombs did not hit in our immediate vicinity. When the naval shells came over, they went directly over our heads and landed beyond us. Some of the bombs though came a lot closer than that.

You know, now when I think about Guadalcanal, I remember it as a large place. When I was on the island, I had the feeling that I could go in almost any direction and not see too many people. When I was there it didn't seem too crowded. There was open space between our tents and the road and on either side. But again, that was a long time ago.

At the beginning of our tour on the Canal, we would always hear sporadic small arms fire and of course artillery—naval shells and bombs. One of the noises I remember well was that of one of our .30 caliber machine guns firing, then being answered by a smaller .25 caliber machine gun, one of theirs. It was like a duel until one or the other stopped firing. When we finally heard only the .30 cal., we knew we had won that one. Sometimes, though, it was the other way around. After a couple of weeks or so, most of the sounds of firing came more from forward positions and didn't really seem to be part of the everyday noise of the place. Our artillery in the perimeter we always heard, especially when we were bombarding the Jap positions.

If there was an air raid on Henderson Field, we would hear the bombs going off, but much of the constant noise was absorbed by the large amount of vegetation between where the bombs and shells landed and where we generally were. The jungle itself seemed to act as an absorbent for the overwhelming noise. That may be another reason we stayed in our bunks so many nights during air raids, we just thought the bombs were exploding a long way off.

All in a Day's Work of Controlling Malaria

Robert Michel PhM3/c, USN Malaria Control Unit Cactus. Back on the New Hebrides, we were issued backpack sprayers. And that should have been the clue as to what our job was to be on Guadalcanal, to locate stagnant water where the *Anopheles* mosquito was breeding and to spray the water with diesel fuel.

The sprayers consisted of a tank that held five gallons of diesel oil with canvas straps that went over our shoulders to keep it in place. Once strapped on and filled up, the whole rig weighed between 35 and 40 pounds. There was a rubber hose that ran from the bottom of the tank to the hand held sprayer. The sprayers had a lever that worked the pump. This lever, which was about one to two feet long, pointed straight ahead to the front. When we wanted to spray, we squeezed the handle on the sprayer and pumped the lever on the bottom of the tank up and down to shoot out the oil. We could also pump it a few times to build up pressure in the tank before activating the valve with the handle.

NAVY DEPARTMENT COMMUNIQUÉ

No. 107, August 17, 1942
South Pacific

1. It is now possible to issue some details of the attacks and landing operations which have been in progress in the Solomon Islands since the early morning of August 7 (local time).

2. The attacks were a complete surprise to the enemy and 18 of their seaplanes were destroyed before they could get into action.

3. Transport-borne, amphibious forces of the U.S. Marine Corps made several landings on islands in the Guadalcanal-Tulagi area. Vigorous enemy resistance was rapidly overcome and a number of Japanese prisoners were taken. The shore positions have since been developed and are now well established.

4. During these landing operations, cruisers and destroyers were so disposed as to protect our transports and cargo ships as they unloaded troops and equipment. While thus engaged on August 7 and 8, our forces were attacked by enemy land-based aircraft. These attacks were driven off and at least 18 more enemy planes destroyed, while only minor damage was suffered by our forces.

5. During the night of August 8-9 an enemy force of cruisers and destroyers attempted to attack our transports, cargo ships and supporting forces. This enemy force was intercepted and engaged by our cruisers and destroyers. The heavy fighting which followed resulted in the enemy being forced to retreat before reaching the vessels engaged in the landing operations. The close-

range fighting during this night engagement resulted in damage both to the enemy and to our forces. The night action is the only engagement between surface forces which has been fought to date in the Solomon Islands.

6. It is impossible, in night engagements, to determine accurately the damage inflicted on the opposing force. No further statement is made at this time of the extent of

Mosquito larvae are visible to the naked eye. They wiggle up and down in the water. Thus, when they come up for air, if a sheen of oil is on top of the water they can't breathe and therefore, die. We also learned how to use dynamite to create drainage ditches to the ocean.

Ours was by no means a glamorous job. It was hot, dirty work. One of the biggest problems we had was that the tanks always leaked. Diesel oil would run down our backs soaking our shirts and our pants. Our clothes and shoes were always damp and our skin was oily and after a little while, it really burned the skin.

I'm not really sure even to this day, but I think we were the only Malaria Control Unit active at that time in the war. Of course that makes sense, the Canal was the only island we were on at that time. Come to think of it, I think some of our group was over at Tulagi also.

The statement by Bob that he thought MCU Cactus was the only Malaria Control Unit active in the islands at that time is not necessarily true and a little clarification is in order. Remember, Bob and Dave had been recruited for malaria control duty back at Noumea, New Caledonia, where they had been stationed with MOB5. When the President Monroe deposited Bob and Dave in the harbor of Noumea in June 1942, as the reader will recall, there were already a number of Navy, Marine, and Army units on the island who had been stationed there for months. The earliest arrivals had, in the words of MSgt. Bill McLaughlin, U.S. Army, "been thrown in the path of the Japanese advance like a log in the road" and had started setting up camp as early as March.

The SeaBees had constructed an airbase at Tantouta which had been active for months, flying men and supplies between Fiji, the New Hebrides, New Zealand, and Australia. Within just three short weeks, Tantouta airfield would be key to keeping supplies flowing to the Marines on Guadalcanal.

The Navy, however, had larger plans for their Malaria Control Units than just the few men in the Cactus group they had just dispatched to Guadalcanal. In June 1942 they already had small units operating at Tontautu airfield in Noumea, New Caledonia, at Efate in the New Hebrides that August, and at the airfield on Fiji just down the road from the town of Nadi (pronounced Nandi) on the island of Viti Levu.

It might be more accurate, then, to say that Bob, Dave and Malaria Control Unit Cactus was the only malaria control unit attached to the Marines, at that time, and required to perform their duties under the daily stress of combat.

If there were no *Anopheles* mosquitoes on New Caledonia and Fiji, then why were there active Malaria Control Units on these islands?

One of LtCmdr. James Sapero's first orders of business was to initiate an aggressive malaria control and eradication program on both New Caledonia and Fiji. Sapero, however, wasn't at all worried about the Anopheles that didn't exist on either island. He was, in fact, more worried that the hundreds of flights returning daily from the combat zones might deliver errant *Anopheles* stowed away either on the planes or on the crewmen themselves. It was the fear of infestation from all the other islands in the area (like Guadalcanal) which were heavily infested by the Anopheles that posed a larger threat to the health of the men stationed on New Caledonia and Fiji—supply bases vital to the continued war effort.

Sapero thought, though incorrectly at the time, that the *Anopheles* could fly a distance of one mile. His malaria control teams, therefore, were ordered set up a perimeter around all the airfields one mile in radius. Navy Malaria Control Units were assigned the task of fumigating every plane that flew into New Caledonia and Fiji as well as every person who stepped off. In addition, any *Anopheles* that might slip through the fumigation process and make it off one of the planes alive would be stopped dead in its tracks within the kill zone set up by his control teams, who were liberally dispensing No.2 diesel oil on every piece of open water within their reach. If, on the outside chance any mosquito made it past both the fumigation teams and the control teams waiting in the kill zone, they would be able to escape no farther than the mere one mile they could fly.

Thomas Gross, US Army 97th Field Artillery Battalion, Americal Division It was my understanding there wasn't a lot of malaria back on New Caledonia where I still was, so I never really heard of any Army Malaria Control Units at that time. We were told many times that there were certain trees there that the *Anopheles* mosquito just couldn't live with.

I went to a local book store just recently and asked the lady that worked there where I could find a book on trees. She asked me what tree I was interested in and I told her a tree that grows in New Caledonia. Lo and behold, she spelled it correctly, and there it was. "Eucalyptus: a tree of the myrtle family having aromatic evergreen leaves."

This is the tree that grew there that kept female mosquitoes, and thus malaria, out of New Caledonia. It was later on Guadalcanal that we had to worry about the *Anopheles*. We were never issued Atabrine on New Caledonia.

Brian Taylor, PhD, Nottingham, UK, Onetime Government Entomologist Malaria Eradication Programme, Solomon Islands. There are no *Anopheline* mosquitoes in New Caledonia, so the eucalyptus trees probably had no great effect on malarial infection since it was spread by the mosquito.

If there was a unit on New Caledonia it would have simply been for mosquito control. Possibly a unit could have been sent to the New Hebrides, which did have *Anopheles farauti*—actually the species was first named from there.

Dr. Taylor introduces an interesting and correct hypothesis regarding eradication and control activities on the islands under discussion. Though there are no malaria carrying mosquitoes present, there was and still is, however, a very aggressive population of the more common annoying, stinging varieties of mosquitoes that incessantly plagued the troops stationed there. So thick were the swarms of these man-eaters, that many men described swarming clouds of mosquitoes so dense that when they landed on a man's back in broad daylight, they resembled a patch of solid black cloth when seen from a distance. Members of the Americal Division, many of whom were stationed on the island for as long as a year before they moved out, tell stories to this day of mosquitoes crawling right around the mosquito bars hung over their cots or swarming up from the ground itself, making their lives absolutely miserable day and night.

The SeaBees

With the Marines barely on the island of Guadalcanal two weeks, Commander Joseph Blundon, 6th Naval Construction Battalion, landed off Lunga Point in a Navy PBY. As soon as he stepped foot on dry ground, he immediately assessed the situation and requested that General Vandegrift order in at least two companies of his SeaBees with heavy construction equipment, maintenance experts, repair specialists, and water purification equipment. Vandegrift agreed and the units were immediately dispatched from Espiritu Santo in the New Hebrides. On his order, the SeaBees joined the struggle.

Its ranks, filled with recruits who were expert engineers and construction tradesmen, the SeaBees, as a group, exceeded the average age of the Marines on the island by many years. It was not uncommon to hear teenage Marines rib the 30 and 35-year-old SeaBees with quips of "Old Man" and "Pop," always, however, delivered with respect.

Nicknamed the "Confused Bastards," a take-off from their units' designation, "Construction Battalions," it didn't take long for the SeaBees to be appreciated by the Marines. Working tirelessly on new construction, fortifications and most importantly expansion of Henderson Field, survival on the island became a direct result of how fast the SeaBees could keep the bomb craters and shell holes in the runway filled in after the daily bombings by the Japanese.

By the time Guadalcanal was secured in February 1943, over 3,600 men comprising the 1st, 6th, 18th, and 26th Naval Construction Battalions had lived, worked, bled, and died alongside the Marines.

Robert Michel PhM3/c, USN Malaria Control Unit Cactus. We didn't have any contact with other branches of the service except the Marines and the SeaBees. The SeaBees were the greatest—the 6th Naval Construction Battalion. They also had the best mess on the island and I do remember we would eat there whenever possible. If you ever needed something done, they could and would do it.

As our unit developed a certain pride and because we were so unique, one of the SeaBees drew up a logo for our unit. The insignia he made depicted a mosquito in flight with the words "Skeeter Beater" or some such over the picture. I can't give those guys enough praise.

Marine Sergeant, Section Leader 75mm Howitzers I/3/11. You know what the SeaBees used to say, don't you? "When a Marine goes to heaven, he'll get there on a road built by the SeaBees."

Robert Michel, PhM3/c USN Malaria Control Unit Cactus. The SeaBees sure were innovative. I remember seeing one of the old Springfield '03s that had been modified by one of those guys for a Marine. He took off the metal floor plate at the bottom of the magazine and then welded on a BAR clip. The BAR clip held twenty rounds and along with the five rounds that the Springfield magazine held, the Marine who owned that rifle now packed twenty-five rounds. The idea was for him to increase his firepower and confuse the Japs at the same time. The SeaBees were truly ingenious.

Day To Day

William C. Dickman, MSgt. USMC Enlisted Navigator/Naval Aviation Observer VMJ-235 MAG-25, South Pacific Combat Air Transport, SCAT. We immediately started flying into the Canal in September after we had been briefed that the Marines there were in a real pickle because the Navy had immediately departed the area after landing them on the island. The Marines were very short on ammo, gas, bombs, food, and mosquito nets. They also had no way to get the wounded out of there.

I still have my logs and, in checking, our first flight into Henderson Field was on September 1, 1942. Do you know what our plane was loaded with that day? Mosquito nets! The whole plane was filled with a cargo of mosquito nets.

Marine Sergeant, Section Leader 75mm Pack Howitzers I/3/11. Mosquito nets? Well, if you tell me that's what that plane carried, I believe you, but I will say, I never saw a mosquito net on the Canal. I guess now I just have to wonder whatever happened to them. I wonder where they ended up—in a pile of trash at the airfield?

Robert Michel, PhM3/c, USN Malaria Control Unit Cactus. I never saw a load of mosquito nets come in at Henderson or anywhere else. We never had nets until a lot later. I wonder who got those, some officers with Division CP?

We lived in pyramidal tents that held eight cots, two on each side. As I remember it, we only had four men to a tent. The doctors had their own tent right next to ours facing the coral road. Our entire area consisted of the space occupied by three of these tents. We kept our equipment in our tents and the drums of diesel oil lined up outside.

Marine Sergeant, Section Leader 75mm Pack Howitzers I/3/11. I never lived in a tent the whole time I was on the Canal. All I ever had was a hole that I dug for myself. I would stack sandbags up around the edges and one or two of us would just lay in there; that's all we ever had. It was especially uncomfortable when it rained because the hole would fill up with water, which was most of the time.

Robert Michel, PhM3/c USN Malaria Control Unit Cactus. As an enlisted man, I didn't have a lot of contact with the doctors, who were officers, in our unit. I do, however, remember one of the doctors, an entomologist, who was a prof at Northwestern University in Chicago. He wanted me to go to school there after the war. His plan was for me to live with his family and be their "house boy," mow the lawn, and do a few odd jobs and chores to pay for my room and board. He would get me admitted to the school and would tutor me. I thought this was a good idea except that I hadn't even been graduated from high school, so how could I attend college without a diploma? Plus, I still had a mother and brother who were dependent on me. As it turned out, I never did take him up on the offer but I have always remembered it.

We were really a loose group, almost no military regimen connected with us at all. In fact, we had no military duties to perform other than our daily job of mosquito eradication, so we got up whenever someone's reveille sounded, put on our cutoff khakis and got some breakfast at the mess we were currently using. We would then return to our tents to fill our backpack sprayers with diesel fuel and then went by jeep or other form of transportation looking for breeding places. When we found them, we sprayed them with diesel fuel and moved on to the next one. We were each given an area to cover and when we were finished, we moved on to another one. On the Canal, we would limit our work to areas where troops were located. We were told once the island was secure, we could then be more aggressive and try to drain water into the sea by using dynamite to create drainage ditches.

Depending on where we were and how much food was available, we would get something to eat about noon at the mess we were near and then continue on until we were finished with that area or until it was time to come back to our base. Then we had to try to find a place to wash off the oil and dirt before calling it a day.

Our main job was to eradicate mosquitoes, not the treatment of malaria. So, we did not treat anyone for malaria, that was done at the hospital or in the field. The hospital, by the way, was nothing more than a couple of tents called the hospital.

We traveled a pretty good distance locating some of those places to spray but not necessarily toward the Jap lines. Most often it was back away from the action where more people were located and around the Henderson Field area where the pilots lived. We never went up to the line to spray. When we first got there, we worked in the immediate vicinity of our camp, then we kept expanding out until we would travel maybe a couple of miles to an area to spray.

I remember that the jungle on the Canal was hot and humid. When we were in the bush, we sometimes had to use machetes to get through the heavy growth. Whenever we were in bush like this, I always had the feeling some damn Jap was up in a coconut tree watching me, which of course, didn't turn out to be true, or maybe he just didn't want to be bothered with one man.

The mud I will always remember. It rained a good bit of the time on the Canal and that resulted in a lot of mud everywhere. That was one of the reasons the guys in the rear areas tried to get a wooden platform built a foot or so off the ground to set their tents on. If you couldn't get a platform the floor in the tent was dirt and would get pretty muddy when it rained. During a heavy rain, you ended up with a floor of mud in the tent. The sides of the tents could be propped up to allow the air to circulate, but unless it stopped raining the floor was always wet, at best.

When it rained, which was most of the time, our shoes were always wet, even when we put them on in the morning. In the morning, first thing, we would always check the inside of our shoes to make sure there were no unfriendly animals inside. Shoes that were constantly wet just helped the jungle rot to flourish.

Marine Sergeant, Section Leader 75mm Pack Howitzers I/ 3/11. It rained constantly and there was mud everywhere. The only tents I ever saw were a few by the hospital. Everything was so wet your clothes would rot and fall right off your body. Our feet were always wet.

William Ricciardi, MM2/c Boat Gang USS Crescent City. It rained there three times a day, and that was during the dry season. During the rainy season, it just never stopped raining.

Robert Michel, PhM3/c Malaria Control Unit Cactus. The jungle rot was a lot like athletes foot, only much worse. It was a deterioration of the tissue of the feet caused by the constant moisture. Literally, some mornings when you put your shoes on, they were as wet as they were when you took them off the night before. The Marines up on the line never took their shoes off and I am quite sure they never even had a change of socks. There wasn't much you could do about it except keep changing your socks if you had them, and keep your shoes dry—both impossible under the conditions we lived in. Once in a while we were able to scrounge up some powder—not foot powder—more like a talcum or baby powder that would help a little. That became a scarce, high priority item.

On Guadalcanal, we didn't have a wooden platform for our tents but we sure did later on Tulagi when we got there. That was paradise. We could scrape the mud off our shoes on the steps going up into the tent. We were even able to use a broom to clean the floor and it was DRY.

I couldn't really say how much it actually rained while I was on Guadalcanal but I do remember it was a lot and it was often!

Brian Taylor, PhD, Nottingham, UK. Onetime Government Entomologist Malaria Eradication Programme, Solomon Islands. On climate—the annual mean temperature in Honiara, for example, on north Guadalcanal is around 27° C, with an annual rainfall of some 3,000 to 3,600 mm.

Robert Michel, PhM3/c USN Malaria Control Unit Cactus. Occasionally, we would go with a Marine patrol taking water and food up the line so we could dispense Atabrine, a little yellow pill that was supposed to make you immune to malaria. It also made you look the color of a lemon. Half the guys refused to take them because they thought the pills would make them impotent, so we sometimes had to stand over them 'til they swallowed.

Naval medical personnel with the Marines on Guadalcanal first began dispensing Atabrine as a suppressive treatment for malaria on September 10, 1942. Atabrine, which is synthesized quinine, is the common trade name for the malaria drug quinicrine. It was developed for the treatment of malaria as well as assorted other parasitical infestations.

Brian Taylor, PhD, Nottingham, UK, Onetime Government Entomologist Malaria Eradication Programme, Solomon Islands. The official WHO (World Health Organization) literature gives the proprietary names as Atebrin, Atabrine and Quinacrine, with the nonproprietary names quinacrine in the USA and mepacrine in the UK.

Atabrine soon became popular in that it had been proven to be a less toxic treatment than quinine. Atabrine's most noticeable side effect, however, was that it imparted a bright yellow color to a patient's urine and skin. Some recipients of Atabrine also suffered side effects ranging from the annoying to the more debilitating such as headaches, nausea, vomiting, abdominal cramping, dizziness, and skin rashes.

Each man on Guadalcanal was to be issued a total of four tablets each week. Those that suffered a toxic or side effect from the Atabrine were to be given quinine in its place. Troops, wary of rumors that it caused impotence and the fact that it would turn the skin bright yellow, the color of a fresh lemon, for the most part refused to take the little pills. Medical personnel were ordered to stand in the chow lines to dispense the drug and to make sure the men swallowed the pills. Atabrine was probably the only effective prophylactic measure mounted against malaria during the campaign and has been credited by medical authorities as actually saving the military situation on Guadalcanal.

John Francis Richter, PhM3/c Corpsman USN, 75mm Gun Half-Track, Headquarters & Service Battery, 11th Marines. Being the Navy Corpsman assigned to a 75mm Gun Half-Track, crewed by five rugged Marines, part of my duty was seeing the men received their proper doses of Atabrine tablets.

Suppressive treatment in the form of 1-1/2 grains of Atabrine was given twice during the day on two days per week. This was administered as a preventative against malaria symptoms.

After popping these greatly hated pills—the men believed they would make them sterile and impotent!!—into their ordered "open wide" mouths, I had to inspect the results. Aided by tongue depressors, I found them tucked under the upper lip, buried under an unyielding tongue, or stuck securely to the roof of the mouth—rather than on their carefree way to the "victim's" stomach.

Notwithstanding my rigid scanning, the grounds around our Half-Track soon wound up peppered generously with those hated pills—fired from the mouths of those extra resourceful Marines who managed to slip through my tight security net.

Marine Sergeant, Section Leader 75mm Pack Howitzers I/3/11. I took the Atabrine every day. I turned yellow like everybody else. I never got malaria the whole time I was in the Pacific but I did get amebic dysentery. When I got home, I was so proud of what I thought was my tropical tan but everybody looked at me and thought I looked like a Jap I was so yellow. I had no idea.

A lot of the guys wouldn't take the Atabrine. They were given the pills and would just spit them right back out. Corpsmen would come up to the lines in the forward positions to make the guys take it, but many still wouldn't.

Our Marine Sergeant neither knew nor had any control over the fact that his recent travel to a tropical country with poor sanitation and unpurified water put him at high risk to develop amebic dysentery, often referred to today as intestinal amebiasis. Those who stayed for any period of time on Guadalcanal along with our sergeant were highly susceptible to an infection caused by a tiny amoebic parasite that attacked the large intestine. The longer one stayed on the island, the more malnourished one became and, therefore, the more prone one became to amebic dysentery.

Once ingested, the tiny parasite could live in the large intestine for long periods of time, frequently without causing disease. Often, however, the parasite attacked the wall of the colon causing acute dysentery and painful diarrhea like that suffered by the sergeant and his buddies on the front lines.

Transmitted by ingesting cysts found in fecally contaminated water or food supplies, and many times assisted in its spread by flies and cockroaches, troops on the Canal had to weather debilitating symptoms that included frequent diarrheal bowel movements accompanied by gut wrenching cramps. The diarrhea, often occurring as many as twenty times a day, frequently contained streaks of blood and mucus. Add to this the nightly bombings, daily shelling, frequent attacks by Japanese snipers and infiltrators, taking your turn on daily patrols, and a two-week bout with a case of amebic dysentery was only that much more pleasant to endure. For those unlucky enough not to get proper treatment immediately, the disease could spread to the liver causing permanent damage or death.

Robert Michel, PhM3/c Malaria Control Unit Cactus. I don't know who it was who declared the Japs a barbaric enemy except that I think it came out of ComSoPac early in the war, probably as a result of the Bataan Death March. As I stated earlier on, we received weapons as a result.

I never did encounter any live Japs while I was on the island, except some prisoners—though I did see lots of dead ones. The local natives hated them and loved us. I certainly feared the Japs. We were told that the Japs on the island were Imperial Naval Marines, elite and fearless.

Many times Japs were able to infiltrate our lines and get back into secondary areas where we were before they were found and done away with. We were always afraid of running into some of them while we were working in the bush. We always carried our .45s with us and at night hung them on the center pole in our tent.

John Francis Richter, PhM3/c Corpsman USN, 75mm Gun Half-Track, Headquarters & Service Battery, 11th Marines. On the Canal I wore a belt that held a holster and a .45 caliber semi-automatic pistol with two additional clips of ammo plus two canteens filled with drinking water—one for my personal use and the other for patient use. I carried two medically filled canvas bags with straps that were harnessed over my shoulders. Later on, I transferred my medical equipment from the bags into a captured and emptied Japanese gas mask case. This made for extra space and easier spotting of needed items.

Dave Minton PhM3/c, USN Malaria Control Unit Cactus. Once in a while, we'd see the occasional Jap prisoner, and of course, a lot of dead ones lying around. Other than that, I didn't see a lot of Japs there. Later, on the ship going home, we had a whole hospital ward full of wounded Jap prisoners, but that was much later.

Patrols

Daily patrols were the lifeblood of the 1st Division Marines those first few weeks on the island. Short a sufficient number of troops to construct and defend a proper defensive perimeter, General Vandegrift was forced to keep his troops on the move covering the gaping holes in his lines. Find the enemy and keep tabs on his whereabouts always topped the orders of the day.

Surviving daily patrols became a very personal experience as the men tripped and hacked their way through the dense, tangled undergrowth of the jungle. Blindly they pushed their way through huge fields of six foot high kunai grass and waded sometimes shoulder deep through stagnant, murky lagoons and swamps. The constant tension of keeping one eye out for the enemy while sweating out the oppressive humidity and unbearable tropical heat took its toll. Dehydration and thirst chipped away at the men's stamina while the threat of deadly snakes, insects, crocodiles, and disease lurked in every creek and ravine.

PFC Stewart Winterton, Marine Rifleman B/1/7. Every day an outfit would go out there somewhere to find out where the enemy was. The marching itself wasn't so bad for us. On Samoa we had been on thirty and thirty-five mile hikes as training for this stuff. The only difference between Samoa and here was, on Samoa a lot of Marines would fall out of formation on the long marches. On Guadalcanal, nobody, but nobody, ever fell out of formation marching.

The humidity and the heat were oppressive. The jungle we went through was very difficult. We always carried machetes that we used to hack through the bush. Following after the first guy, though, the trail would get to be pretty much OK to move over.

Marine Sergeant, Section Leader 75mm Pack Howitzers I/3/11. When we were in position out in the area that would later become Fighter One, we had tall kunai grass to our front and solid jungle behind us. The only patrols we would run were ones sent out into the jungle behind us to make sure the area back there was clear of Japs. The jungle was so thick that no large units could travel through there but individuals could. The jungle would stop any large units, so we were just making sure no small groups or snipers were moving around back there.

Robert Michel, PhM3/c, USN Malaria Control Unit Cactus. I don't really have any idea why we went on patrol with the Marines other than to take the Atabrine up to the men in the forward positions. I rather think that many times it was because of boredom with what we were doing every day. There was only so much stagnant water we could cover with diesel oil without going nuts. Rather ironic that we should choose to go into harm's way with the Marines just to shake off a little boredom.

Dave Minton PhM3/c, USN Malaria Control Unit Cactus. I don't think we went on patrols with the Marines out of boredom. I think it was more to get into an area we needed to take a look at or evaluate. If we needed to see an area that we could only get to by going out with the Marines, then we would go with them.

Marine Sergeant, Section Leader 75mm Pack Howitzers I/3/11. We never really stayed in the same positions all the time. We would move around and set up our guns wherever we may be needed. Our guns had a range that could cover from Kukum on the west all the way to the Ilu River on the east and everything in between.

We were in a position for a while east of Henderson Field with the jungle at our back and a huge field of kunai grass to our front. The kunai grass was about four to five feet tall and very sharp. The Japs were always sneaking around in the grass to see if they could get a look at our positions or to see if they could get into our lines. The Japs used to set off firecrackers and see what they could see. Some of our guys would open up with machine guns when they heard the firecrackers going off, giving away the positions of the guns. The Japs would then move in as close as they could and lob something in at us. This happened more early on when we were still pretty green.

To see where the Japs were, we would send out patrols of five or six guys everyday into the grass in front of our positions. One day we sent our patrol out and one of the guys was a young kid who was a real character. He was on his belly cautiously crawling through the grass with the rest of the patrol spread out around him. He would crawl a little and then part the grass to see what was directly ahead of him and then move forward a little more. Well, he crawls, stops, and very carefully parts the grass. Six inches in front of his face is a Japanese soldier who was doing the same thing. Imagine their surprise. These two guys, six inches apart and staring right into each other's face. They both jumped up off the ground screaming and turned right around and ran back where they had come from.

When the young kid reached our position, he just stopped, suddenly realizing he was supposed to shoot the Jap. He turned around just as quickly as he came running in and started firing back where he saw the Jap. By that time, though, the Jap was long gone.

William Ricciardi, MM2/c Boat Gang USS Crescent City. I made over twenty-five trips into the Canal and ended up left on the beach many times when our ship pulled out. I was also sent ashore if they needed maintenance on the boats at Kukum. The boats there at the boat base were always in bad shape. I was in all the actions there with the Marines—the Matinikau, Point Cruz, Aola Bay, Cape Esperance, and more, even after the island was secured.

There was one Marine I'll never forget who would be waiting for me on the beach when he saw me coming in. He would ask me if I had any of "the good stuff." You know, pogey bait, booze, cigarettes. One time he asked me if I had a machinist's screwdriver. A machinist's screwdriver is real long. I asked him what he wanted a machinist's screwdriver for and told him when I went back to the ship I would steal one for him. The next time in I had the screwdriver and gave it to him. You know what he did with that screwdriver? He sharpened the end of it to a long point like an oversized ice pick. He would sneak out of the lines at night and wait in the bush for a Jap patrol to pass by and then sneak up and grab the last Jap in the line as they went by. He would grab that Jap around the head and twist his neck and stick that screwdriver right up into his sternum. I asked him if he wasn't afraid of being heard. He always said, "When you stick that screwdriver in 'em, they don't even make a sound."

Dave Minton PhM3/c, USN Malaria Control Unit Cactus. There was a big difference between us and the Japs. To them it was OK to die in battle. They believed they would just go to heaven. Our guys didn't think that way. It wouldn't take but one or two Japs with an attitude like that on the Canal, or later on the other islands, to tie up a whole lot of American troops. This really was a problem later in the war on some of those other islands. The Japs didn't ever outnumber us when they were defending those islands, it was just a few that always held us up. You have to understand they had a totally different outlook on life than we did.

Contrary to what people will tell you or what you might read in books, they outfought us every time we got into it with them. The only time we beat them was when we outnumbered them—when we had more planes, more ships, more tanks, more ammunition, and more men. You've heard that before, we beat our enemies in World War II because we out-produced them in every way. If it wasn't for that, they would have beat us.

We say we beat the Japs in a lot of those naval battles down there. We didn't beat them in a single one. The only times we beat them was when we had ten ships and they only had two. Their ships used to really kick the shit out of our navy. They outfought us and they could outshoot us. We finally beat them and the Germans in Europe because we had a lot more equipment than they did. That's how we really won—by outnumbering our enemy. It wasn't until we ran up against the Chinese in Korea that we were ever outnumbered by our enemies. In Korea, those Chinese often had us down by as much as 15 to 1, but that's another story I won't get into right now.

Robert Michel, PhM3/c USN Malaria Control Unit Cactus. It had been drilled into us that the Japanese had no respect for medical personnel, so whenever we went on patrol with the Marines I would always ask for a BAR. I never really "owned" one; rather, it would be supplied to me by whatever group we were going on patrol with. Because of its rapid rate of fire, I would also carry about five or six extra clips of ammunition for it. Each clip held 20 rounds. Besides the gun and the ammunition, I carried my medical kit.

A corpsman's medical kit was a small canvas bag about three times the size of a kid's fanny pack of today. It had a strap on it to sling it over your shoulder. The kit consisted of 1/4 gr. morphine syrettes, bandages and tourniquet, and a lot of other medicines including acetylsalicylic acid (aspirin), APCs (aspirin, phanastine and caffeine), and anything else you might want to put in it. The bandages in the kit were mostly 3" X 3", some 6" X 6" and 8" X 8", all in sterile packs, one per package. There was adhesive tape and rolls of 1" bandages, along with sulfanilimide, sulfathiazole, sulfadiazine and of course, quinine. Not being on the line everyday, I never had an occasion to use the sulfanilamide or the sulfadiazine. The Company Corpsmen used to sprinkle it generously into open wounds to prevent infection. It evidently worked very well.

The sulfathiazole was an oral medication and was used frequently. When I got blood poisoning the following year from a coral cut on my ankle while diving off Florida, I took sulfathiazole. It probably saved my life, or at the very least, my leg. About five years ago, I went down riding my motorcycle and badly burned my leg on the exhaust pipe. It was bad enough that I went to the ER to have it taken care of. They put an ointment and a bandage on it and sent me home. A few days later, it was not only no better, it was worse. I went back to the hospital to have the burn looked at again and they concluded that I was allergic to what they were using, a sulfa drug. Ironic, isn't it? Hell, I believe on the Canal we were the first to even use the sulfas. I think they were the first of what were then referred to as the new miracle drugs.

The medical kit itself was not very heavy. Although, altogether, the BAR, the ammunition, and the rest of the gear was a fairly heavy load. The only time I carried one of those medical kits around was when I thought it might be needed. Given my job, that was not very often.

When a patrol was scheduled, we would form up near the area we were going to start out from. Usually, the gathering point was near some kind of a supply dump or something like that. There was usually a Buck or Staff Sergeant in charge and he would go around the group asking if everyone had what they needed. If I said I wanted a BAR and there was one available in the patrol group, he would generally give it to me. Hell, I was eighteen years old and I knew everything there was to know about anything, but I never even fired a BAR. I was just thinking about that the other day. I don't think today I would even know how to change a BAR from repeat to automatic.

Marine Sergeant, Section Leader 75mm Pack Howitzers I/ 3/11. The BAR was a great weapon, although you had to be trained on it to shoot it. It was a sophisticated weapon and tricky to fire. That was because it could be fired on full automatic. It was a great support weapon to have with the unit, though.

Robert Michel, PhM3/c USN Malaria Control Unit Cactus. Now you might have heard stories about how Marines and Navy guys didn't get along. Well, that was true Stateside before the war started, but overseas on the Canal, there was always a special relationship between the Doc and the Marine. Loaded down the way I would be whenever I was out on a patrol, there would always be one Marine who come up alongside and ask, "Carry your pack, Doc?" It was never really said, but the Marines always liked to be near us Corpsmen just in case they got hit. Of course, we always wanted to be near the Marines in case of trouble. Those guys would never let you down even in the toughest situation. Those 1st Division guys were the best.

Patrols! God, does this one stand out. Dave and I are on this one patrol, just a short time after we arrived on the island. As we passed a clearing on the way up, the word comes down, "Land Mines!" We looked at the clearing and saw a foxhole with a couple of dead Japs inside. All around the clearing were fifteen or twenty round, flat things that resembled an oversized surveyor's tape-measure. Dave picked up one of the Jap rifles and I picked up one of those other things and began "inspecting it" by unscrewing a cap in the center of the thing. I was going to go deeper but my guardian angel said No, so I screwed the cap back on and threw the thing on the ground. Later I was told that after I had unscrewed that pressure cap, had I even touched the inside, I wouldn't be telling this story today. We were sure naïve and green…and lucky.

Dave Minton PhM3/c, USN Malaria Control Unit Cactus. I remember going in there and picking up that Jap rifle, a .25 caliber. Bob picking up the mine? He was always doing some crazy shit like that. I don't know why he did it—like sitting around and disassembling hand grenades, I suppose. He was nuts.

Robert Michel PhM3/c, USN Malaria Control Unit Cactus. One day in September, Dave and I hiked up toward the Matanikau River through an area that was literally covered with dead Japs. The area looked like it had been shelled by either our artillery or the Navy because the tops of the coconut trees had all been blown off and dead bodies were laying all over the place. The stench of so many decaying bodies was unbelievable. The one thing I will always remember about that area was the sight of ants going in the eye sockets of the Japs and coming back out with pieces of tissue in their mouths. The bodies, though, were mostly intact, so the shelling must have happened only a day or two before. It usually took about three or four days for a dead Jap to deteriorate completely in that climate.

Another time, Dave and I walked in a general northwest direction from our base about two or three miles to an area that had obviously been used by the Japs as a hospital of some kind. It was a building, like a long house built by the island natives on stilts, about three or four feet high, about fifty feet long and fifteen to twenty feet wide. It was made of whatever material the natives used for building such dwellings. There were still some bodies lying on mats inside the building.

I heard rumors about the place that some Marines had gone in there and killed everyone, including some Jap women. The story goes that some of the guys cut off the women's breasts to take home as souvenirs. How true the story, I don't know. I do know that there were some Marines who would ask me if I had some formaldehyde or anything to preserve something in. I would usually give them some denatured alcohol. There were a lot of ears that ended up going from the dirty pockets of Marines to a bottle of alcohol. I wonder how many of those bottles ended up on mantle pieces back in the States.

And gold teeth! Some of the Marines would knock out the gold teeth of any dead Jap they found. And as I remember, the Japs all seemed to have a lot of gold teeth. There was one Marine who had a bag full of them. He must have had ten pounds of gold teeth.

I guess there was barbarism on both sides. It must have been the times and the fierce hatred we had for the Japs. I know I hated the Japanese until not too many years ago.

Marine Sergeant, Section Leader 75mm Pack Howitzers I/3/11. We were waiting for motor transport one day to finish moving our guns into a new position and the edge of the jungle was right out in front of us. All of a sudden a Jap sniper started firing at us from out there. One of our sergeants comes up and asks if anyone sees where he is. The Japs used smokeless powder in their ammunition, so when they fired you never knew where they were unless you could actually see them. Our ammunition wasn't smokeless. When we fired a round it was like a Revolutionary War musket going off with this big puff of smoke. That put us at a disadvantage every time we fired our weapons.

Anyway, we were all ducking down and trying to spot the sniper, waiting for motor transport to get us moved, when the sergeant spotted the Jap behind a fallen log. He had a .45 and took aim at the spot where he last saw the Jap, waiting for him to pop his head up again. The next thing you know, we saw the helmet pop up and the sergeant fired. He hits that Jap right on the helmet, but the round didn't penetrate it, it only put a dent right in the middle of it. It stunned the Jap and he fell back behind the log. The sergeant runs up and sees the Jap lying behind the log. He fires six rounds into the Jap, two in the chest, one in the right leg, and one in the left leg, and missed with his last two shots. The Jap was just a young boy like all of us were, and the sergeant reaches down and drags him out by the ankle. The Jap is laying there on the ground and lets out a couple of gurgles and dies.

The lieutenant comes over and tells the sergeant, "OK, you can have all the souvenirs." You know, watches, weapons, money, whatever. Then the lieutenant tells the sergeant, "OK, now you can bury him." The lieutenant didn't want the body lying out there stinking the place up after a couple of days. The sergeant drags the body out about a hundred yards in front of our position and carefully measures the body from his head to his waist. He digs a hole about thirty inches deep and crams the Jap into the hole, only the arms and legs wouldn't fit. He twists and bends them in and then covers the body over, carefully tamping the ground when he was done.

The next day, it rains like hell all day and night. Wouldn't you know, out pops an arm. The lieutenant sends the sergeant out there and has him cover the body again. It keeps raining and out pops a leg. The lieutenant tells the sergeant to get back out there again and bury that Jap. This went on for three days until the body rotted away.

Chapter 6

Conversations on the Reality of War

ALTHOUGH WAR HAS always been the ultimate act of inhumanity, World War II in the Pacific was particularly barbaric for reasons unique to the cultures of the combatants. Foremost, perhaps, was the cultural definition of honor embraced by Japanese society. The Japanese held strong the conviction that the highest level of honor was achieved by dying for their Emperor. Part religious belief and part bushido—the warrior code within which Japanese youth were raised—their troops fought tenaciously to the last man in every engagement with their enemies. Surrender was not an option for either party in the conflict. To the Japanese, surrender brought dishonor to the individual, as well as to their family, for generations to come. It was their firm belief that anyone who surrendered had no honor and was therefore, less than a human being. The Japanese were baffled by the wholesale surrender of Allied forces all across the Pacific in the early days of the war. Captured Allied servicemen were brutally treated by their Japanese conquerors because they had brought dishonor on themselves and were, therefore, entitled to no better treatment.

As the war progressed, U.S. troops, thus locked in a continuous death struggle with an enemy that would never quit, developed a vehement hatred for the Japanese. A Japanese soldier was viewed as a sneaky, animal-like creature not far removed from their simian ancestors. Poor of sight, small of stature, and barbaric in their behavior, the Japanese people as a whole were virulently despised as the yellow bastards who slaughtered unsuspecting American servicemen at Pearl Harbor, starting the war in a most cowardly fashion.

Lock the two sides together on small Pacific islands with no room to maneuver, and it became the ultimate duel to the death. Neither side either gave or expected quarter. Combat was almost always hand-to-hand, resulting in the most inhuman activity. All the most modern weapons of war were employed, and when they were expended or failed, combatants resorted to the use of their bare hands to strangle, gouge, and batter their enemies to death. The closer U.S. forces moved toward the Japanese homeland, the more steadfast the defenders became. U.S. troops knew that no island they attacked would ever be secure as long as one Japanese soldier remained alive, and Marines, in particular, were more than willing to accommodate their Japanese foe.

Marine Sergeant, Section Leader 75mm Pack Howitzers I/3/11. I want to tell you about our outfit's Corpsman. Right after the Battle of the Tenaru, which was really at the Ilu River, we came out of our positions and there were dead Japanese lying all over the place. We were moving through the area and stepping over bodies everywhere.

Our Corpsman spots a Jap, still alive, who was lying on the ground. He was moaning and had a shell fragment stuck right in his forehead. The Corpsman pulls out his pliers and yanks the thing out of the Jap's head and starts fussing over him like he was one of ours. He bandages the Jap up and props him up against a tree.

The next day, this Jap is still alive and the Corpsman is talking to him and consoling him, telling the Jap he is going to be OK and saying, 'Come on, buddy, you're going to make it. You're going to be OK.'

The Jap, who had been comatose up until now, started to come around, regaining consciousness. He looks around, rolling his head, getting a sense of what was going on around him. Focusing on the face in front of him, he realized the person talking to him was a Marine, and with that, leans back and spits right in the Corpsman's face. The Corpsman reels back screaming, 'You ungrateful son of a bitch,' and pulls his .45 and shoots the Jap right in the head.

Tom Croft, BM3/c Gun Loader Turret 1 USS Monssen DD436. If we had a general feeling about the Japanese, it was nothing more than to just kill the SOBs.

William Ricciardi, MM2/c Boat Gang USS Crescent City. You hear people refer to them as Japanese. To me they're Japs and nothing more. When we first started fighting them on the Canal, we had a lot of green kids there. They fought them as good sports, the American way. It didn't take our kids long to figure out they had to fight the Japs the same way they fought us. The Japs had no compassion whatsoever.

There were never a lot of Jap prisoners taken on the Canal, and even fewer as the war went on. Once in a while, though, we would see the Marines throw prisoners in the back of a Jeep and head back to the stockade with them. A lot of times they wouldn't make it, you know what I mean? Another time, I was at the stockade and there were three Jap prisoners in there huddled in the back. The Marine guard opened the gate and waved to the prisoners, motioning for them to go ahead and leave. He turned his back on the prisoners and kept waving for them to just go. One of the Japs in there just shook his head—he knew what would happen to them if they left. He knew what the game was all about. We all hated ‘em.

William C. Dickman, MSgt. USMC Enlisted Navigator/Naval Aviation Observer VMJ-253 MAG-25, South Pacific Combat Air Transport, SCAT. My feelings at the time were not good toward the Japanese. We had not forgotten Pearl Harbor, where several of our buddies were killed. I was also wounded at Pearl Harbor, so my feelings for the Japs were not good at all.

PFC Stewart Winterton, Marine Rifleman, B/1/7. We hated the Japanese, but we never really had that much desire to get into the action as I recall.

Thomas Gross, US Army 97th Field Artillery Battalion, Americal Division. The average Jap was nothing but an ignorant peasant—no drive and no imagination when he was on his own. One American was a far better all-around soldier than a Jap. All we had to do was get at him where he couldn’t climb trees and where he couldn’t dig in and find cover. That was the only way we’d ever get the war over with, and without losing a lot of men. I guess we’d have to wait for the Germans to fold before we could really tackle the Japs in earnest. We couldn’t wait until that day came. We thought, My God, imagine the whole U.S. and British fleets in the Pacific? We’d throw a line of ships around Japan they could never get through.

At the time, I thought I would rather fight the Germans than the Japs, because at least the Germans were men and not rats. I had no qualms about killing a rat back then. I used to think that one doesn’t realize what evil-looking things they were. I thought they didn’t even look human, but I did respect them, at least in the jungle. Oh, how I wished I could have seen them trapped in a country like North Africa.

I used to think back then that the only trouble about killing a Jap was that he stank so bad if he wasn’t buried. From all I saw at the time, no one had the time to bury the Jap dead, not even the Japs. Pleasant people those Japs.

Washing Machine Charlie and Other Nightmares

Robert Michel, PhM3/c, USN Malaria Control Unit Cactus. We were never really an active part of the fighting units, although we were right there during the bombardments from the Jap ships and the bombs from Washing Machine Charlie who flew over us every night to bomb us. Henderson Field was close to where our tents were and was always a prime target for the bombers and the Jap ships.

At the beginning, we would spend most of the night in a foxhole, or covered slit trench. The slit trench we used was about twenty-five or thirty feet away from our tent. I knew exactly where it was in relation to the tent doorway so that if it was pitch dark when I ran out of my tent, I would just have to bear a little to the left to get to the entrance. It was a trench big enough to hold five or six guys and was dug by the SeaBees. The trench was covered over with coconut logs and some kind of roof. It was about four to six feet deep and only about two to three feet wide.

Those trenches were very cramped. To get in, you had to jump down in the open end and crawl down to the far end so there was room enough for everyone; you were all under the log roof. It was the roof that provided cover from the bombs and flying debris during the bombings. Everyone crammed into that covered end of the trench and huddled there until we heard the 'All Clear.' Sometimes we would spend the whole night in there—sometimes only an hour or so.

PFC Stewart Winterton, Marine Rifleman, B/1/7. My first night there, Washing Machine Charlie came over. I did not dig a foxhole that first night because I was too tired. My buddy and I slept on top of the ground. When Washing Machine Charlie dropped two flares, followed by two bombs, you never saw two guys dig so fast. We weren't virgins after that raid.

Every night up there in that area, all you would do is shiver and shake—scared to death. Every single night Washing Machine Charlie came over, so we got used to sleeping one hour off and one hour on watch, hoping like hell the guy in the next foxhole over wasn't sleeping instead of watching.

Lt. Cmdr. Lewis P. May, USNR Ret., Assistant Supply Officer, USS President Hayes APA-20. My captain let me deliver some provisions to Guadalcanal on one of our trips in there, and I stayed overnight with the Marines on shore. Washing Machine Charlie, a little Jap plane, flew over us that night, and although everyone laughed at him, a 'Red Alert' had to be declared and we went into foxholes.

Marine Sergeant, Section Leader 75mm Pack Howitzers I/3/11. I remember there were two different ways the Japanese would bomb us. The Japanese planes used to come over us in a V formation, often with as many as twenty-seven planes in the formation. The V's were made up of little V's of three planes, and then three little V's, nine planes, to make a larger V, and so on to the full formation.

Anyway, the first type of bombing I called 'salvo bombing.' The formation would fly over their target, and all the planes in the formation would drop all of their bombs at one time. The bombs would sometimes hit very close, and the concussion of all those bombs going off at the same time was tremendous.

The second type of bombing I called 'pattern bombing.' When pattern bombing, the Japanese formation would fly over the target and as the first planes in the formation reached the target point, they would release their bombs. The planes behind them would then release their bombs as they reached the target point, so you had the planes dropping their bombs in a continuous flow. I used to hate this type of bombing. It sounded just like a huge giant walking towards you. Boom. Boom. Boom. The explosions sounded like huge footsteps getting closer to you. The bombs would hit close to Henderson Field first, which was always a prime target for the Japanese, and then you would hear them coming towards you from the airfield. I always preferred the salvo bombing—at least the bombs would all go off at once and be over with, whereas with a pattern bombing, you couldn't do anything but sit there while the detonations kept moving closer.

William Ricciardi, MM2/c Boat Gang USS Crescent City. It was no easy job getting up to the Canal from New Caledonia with supplies. It got to the point that every time we came around a point of land on the Canal we called Point X-ray, there was always a Jap bomber waiting there for us. He would let go a stick of bombs that would luckily miss us. One time, though, one of the bombs fell so close to where I was standing next to a bulwark on the bow of the ship that I couldn't see anything because of the column of water it kicked up. I was knocked flat, and you would have thought someone would have checked to see if I was OK after the water settled and the smoke cleared.

Robert Michel, PhM3/c USN Malaria Control Unit Cactus. It got to the point, though, that when Charlie came over, there was a distinctive sound to its engines. When he started dropping his load, we took the attitude, Hell, if I'm going to get it, at least I'll be asleep. Instead of going to the foxholes, we'd roll over and go back to sleep. Strange, isn't it? We valued a good night's sleep over life itself. Not so when the naval shells came in. The sound of one of those naval shells going overhead had a different sound than that of a bomb coming down, and when it exploded it sounded totally different from a bomb going off. Shelling was really scary. When I look back on it, I still get the chills.

Marine Sergeant, Section Leader 75mm Pack Howitzers I/3/11. At night, our sentries would stand out there and watch the horizon over the channel. They'd call out, 'Flash' when they saw the explosion of a naval gun go off. The silhouette could be seen for miles. First we would see the flash, then the shell would come over, and finally we would hear the sound. The shells traveled faster than the speed of sound, so the sound was always the last thing we heard. By the time you heard the shell, it was usually too late. Sometimes, too, the flash on the horizon would be nothing more than lightning from a tropical storm, which was very frequent. In that case, the sentries were only calling out a false alarm. It was hard to distinguish the real thing from the lightning sometimes.

William Ricciardi, MM2/c Boat Gang USS Crescent City. We used to make runs across the channel from the Lunga boat base to Tulagi to the Combat Engine Repair Shop that was built over there. Or sometimes, we'd go just to pick up supplies and

"Pistol Pete," Captured from Japanese
Guadalcanal, 1942

stuff. It was a twenty-five mile run over there and I will never forget how watchful we had to be for Jap subs. They would surface and start shooting at us if they saw us out there in the channel. One time, a sub came up and started shooting, and would you believe they were so cocky, the crew was actually taking turns at the gun. We saw the crew on deck taking turns shooting at us…unbelievable.

Robert Michel PhM3/c, USN Malaria Control Unit Cactus. There was also Pistol Pete, as we called him. The Japs had this artillery piece that they would use to lob shells into our general area from time to time. I guess they used it to try to not only kill some of us but to also demoralize us at the same time. We never knew when it would happen, but every day or so we would hear these loud shots being fired and knew it was Pistol Pete in action again. As the Marines defeated more Japs, they also captured Pistol Pete, and that ended that shelling.

Just prior to the Japanese naval bombardment of Henderson Field on October 13, enemy troops landing by night west of the Matanikau River brought with them the first long-range artillery ever seen on the island. Able to lob shells at will into Marine positions within the tiny perimeter, the Japanese gun positions were well camouflaged and beyond the effective range of any response the Marines might mount.

Victims of the random and unseen artillery attacks were quick to dub the offending Japanese artillery, Pistol Pete. Hence, when any artillery shell fell within the perimeter, it was said to have originated from the offending barrel of the one and only Pistol Pete. Never escalating beyond the level of an annoyance, because the Japanese lacked an abundance of artillery ammunition for the big guns, schemes to locate and destroy the elusive Pistol Pete would occupy the foremost military minds of the Marines for many weeks to come.

Robert Michel PhM3/c, USN Malaria Control Unit Cactus. The bombings, though, occurred every night without fail, until one day one of the guys in our bivouac area went bananas and eventually had to be evacuated. It was after an air raid that we found this guy still in the far end of the foxhole under the log roof not wanting to come out. The biggest problem with this situation, other than the threat to his own well being, was that he had his rifle with him. He was convinced that anyone trying to enter the bunker was a Jap. He would fire off a round or two whenever he saw or heard anyone trying to get in through the entrance. We would each take turns going up to the entrance and ask him how he was doing. This went on for two days before we were able to talk him out of that bunker.

About a month later, a similar thing happened all over again when another guy cracked up. This time, unlike the previous incident, the guy was a doctor and was in his tent. When I went into his tent to talk with him, he was incoherent. He was out of it. His conversation with me made no sense at all. I went back outside and talked to some of the other guys who were milling around and told them what I thought had happened.

I am not so sure that we were all suffering from some form of battle fatigue at one time or another. Maybe that was what was happening to us when we said, 'The hell with it during the air raids,' and just rolled over and went back to sleep. These guys, though, were the first to outwardly lose it, and we weren't exactly sure how to react or treat them.

After all these years, the one feeling I remember having when these guys cracked up was one of absolute sadness and relief that it wasn't me. I know I felt sad that they had gotten to the state they were in mentally. It was tough on all of us.

Yet with everything that was going on all around us, we just went on with our daily routines without too much more thought about them. I can only think it was the day in and day out tension that caused so many men to move into a different world. I only saw this happen to the two guys while I was on Guadalcanal, but still feel a real sadness with the memories. I also feel a sense of relief that it didn't happen to me.

Battle exhaustion, combat fatigue, shell shock, combat neurosis, LMF (lacking moral fiber), and malingering were all terms used with frequency by officers and medical professionals alike throughout the war to describe the curious psychiatric reactions of men living under the constant strain and traumatic shock of combat. For the Marines, aviators, and sailors living in squalid conditions on Guadalcanal, shelling, sniping, and bombing were their daily fare. Every afternoon, like clockwork, between noon and 2:00 P.M., Japanese bombers visited the perimeter on their daily runs from Rabaul. Every night, Washing Machine Charlie sputtered over the island dropping a bomb or two at random just to keep the troops awake. Mix in the constant threat of never knowing for sure where the enemy was, lack of food, the imminent threat of unknown disease, and the ever-increasing factor of time, and Guadalcanal was a stress load only a few unique individuals were capable of carrying.

First diagnosed in World War I, shell shock as defined in World War II medical terms, was just one more malady to be suffered by the troops. At first thought to be an actual physical condition caused by blast and carbon monoxide poisoning, it was later realized that the symptoms of mental confusion and terror were actually neurological in nature. Combat fatigue, it was realized, had similar symptoms to those exhibited by victims of traumatic accidents such as train wrecks, fires, and explosions. Rest, away from the cause of the terror, usually returned the victim to normalcy in a short period of time. At Guadalcanal, this was impossible because it meant rest away from the front lines, where every man was needed. Some of those that suffered more severely, however, never fully recovered, carrying with them more serious side effects such as nightmares, flashbacks, paralysis, or loss of sight, speech or hearing. Often times it left the victim with a pronounced limp. Today we hear the term Post Traumatic Stress Disorder attached to disorders suffered by Vietnam veterans. It could also be used to more precisely define the malady suffered by thousands on the battlefields of World War II.

We will never know for sure how many suffered the horror of shell shock on Guadalcanal and how many still carry the scars of its effects yet today. Perhaps when Bob says they could all have been suffering from some sort of battle fatigue at one time or another, he is accurately describing the mental condition of many of the combatants on Guadalcanal in more realistic terms.

Brian Taylor, PhD, Nottingham, UK, Onetime Government Entomologist Malaria Eradication Programme, Solomon Islands. I remember reading a confidential document in the Medical Department records that dated from the days of the battles on Guadalcanal. If I remember rightly, apart from the death rate from malaria, the second highest cause of U.S. Marine deaths was given as combat, but third was suicide and fourth was "accidents," with both the latter being of, to me, a surprisingly high level. In 1971, our programme was fortunate to have four U.S. Peace Corps volunteers assigned to help us. One, named Ken, had been a medical corpsman in Vietnam, suffering leg injuries when the man ahead of him stepped on a landmine and was killed. When I asked Ken of suicides in the Vietnam combatees, he said there was extensive use of tranquillizers—perhaps a reason the military action was not as good as it might have been?

Robert Michel PhM3/c, USN Malaria Control Unit Cactus. I have a vague recollection that the two psychos were replaced after that. Other than these two guys who went ape, our command changed little, including our day-to-day routine. None of us was ever really seriously injured, except for the malaria, dengue fever, jungle rot, and dysentery we all got. I'm not so sure the dysentery wasn't the worst. That really put you out of business.

You never really had any warning. All of a sudden, one day, you just had it. We had a three-holer near us, and sometimes there was a line, figuratively speaking, to get in. The cramps were indescribable. If you had dysentery along with dengue or malaria at the same time, you were in your own private hell. With the dysentery, your bowels could be empty and you would still have the intense urge to go. It was brutal. Sometimes the dysentery would last three or four days, sometimes longer.

Marine Sergeant, Section Leader 75mm Pack Howitzers, I/3/11. The Japs used to use what we called 'daisy cutter' bombs on us. When they landed, they would only sink into the ground about six inches and then they exploded, throwing shrapnel out in a complete three hundred and sixty degree circle around it. The concussion from one of those things was really tremendous when it went off.

One of our guys was hiding in a shallow slit trench off by himself a ways from where all the rest of us were. We were getting hit, and one of those daisy cutters landed about eighteen feet from this guy's hole. The explosion was huge. This guy gets up out of the hole and starts wandering around speaking gibberish. The concussion was so huge it caused him to lose his hearing. Before long, he starts reciting passages from the Bible. He became a real "Bible thumper," and we all kidded him about it. Later on, when we got to Australia, human nature got the better of all of us, including him, and we all reverted back to our old behavior.

Other than that, all of the guys in our outfit stood up real well to everything thrown at us.

Our Daily Bread

Robert Michel, PhM3/c, USN Malaria Control Unit Cactus. The food on Guadalcanal was always in short supply, and we kept our water in a Lister bag that we always had nearby to fill our canteens from. The water always tasted terrible. I guess they poured iodine or chlorine in it. There were only a few times when the bag was empty and we had to really look around for water. I can only imagine that those guys on the line had it pretty bad in a lot of ways, including lack of water.

PFC Stewart Winterton, Marine Rifleman, B/1/7 Water was always in short supply for us, but with all the creeks and lagoons around and two drops of iodine, you could drink almost anything. One day I was really thirsty and filled my canteen in a lagoon and looked over and saw two rotting Japanese corpses in the water. Didn't bat an eye, I put the two drops of iodine in my canteen and drank away.

Marine Sergeant, Section Leader 75mm Pack Howitzers, I/3/11. No, I don't remember water being a problem for us. The Lunga River was a fast flowing river and a good supply. Unless we were out in an area where there were no rivers or streams nearby, we didn't really worry that much about water. By the Lunga River there, the Japs had built a big bridge; they even had an ice making factory there.

Robert Michel PhM3/c, USN Malaria Control Unit Cactus. Lack of food, however, was critical most of the time. Often, there just wasn't any at all. I know that sometimes we would go to the beach and catch some of the land crabs—weird things that scurried around the place mostly near the water. We would cook them over an open fire and then eat them. As I remember, they didn't taste bad. They just looked horrible. I remember one time we were near the water looking for some land crabs when I spotted what looked like a melon growing in the sand. I picked it up and cut it open and sure enough, it was a round watermelon. How it got there, the Lord only knows, but it sure tasted good. We must have been the only guys on the Canal who had watermelon.

Sometimes one of the cooks would make a deal with someone aboard a ship off shore for some food, and it was like a feast for us. Otherwise, if we were lucky, it was Spam and powdered eggs and bread with canned butter. I remember reading an advertisement in a newspaper or a magazine that found its way ashore at the time that said the butter sent to U.S. troops in the South Pacific wouldn't melt at 100 degrees. They were right, it wouldn't. And when you consider the body's temperature is 98.6 degrees, it was like eating axle grease. Rather than feed it to us on bread, the cooks would use it to fry food and give us jam, if they had it, for the bread. The cooks must have had a good supply of flour and yeast because we had fresh baked bread from time to time.

We would also once in a while get some of the Aussie rations—their tinned beef, especially. I had heard of their canned sheep's tongue, but I don't remember ever having any of that. Today, however, whenever I can find it in the stores, I buy it. Here they call it lamb's tongue, just like pickled pig's feet.

The coconuts were pretty good. I learned later on the Floridas from the natives, though, that nobody eats a ripe coconut. You open it when it is green, drink the milk, and scoop out the meat with a spoon or your thumbnail. The meat is very soft and only about 1/8 thick. Totally different from what you would buy in the store. Drink or eat too much coconut, and you end up with the runs. The bananas were also different. They were short, stumpy, and hard. The natives would cook them rather than eat them fresh.

Marine Sergeant, Section Leader 75mm Pack Howitzers, I/3/11. We used to eat the coconuts and that was OK, but if you drank the milk, then you got the runs and that was the end of it.

Brian Taylor, PhD, Nottingham, UK, Onetime Government Employee Malaria Eradication Programme, Solomon Islands. I find this an unlikely story as my wife, who grew up in India, always swore by green coconut water, dap as she calls it, as the best thing for anyone with diarrhea. Personally, I drank it often when out on field tours.

One memorable day on the beach at Tuhugago, Rennell Island, where there was no source of fresh water, the dap was a very important factor as the expected ship was delayed for over 24 hours. Nowhere have I ever suffered from drinking dap and it is an excellent thing to give to small babies when other water quality is dubious.

Thomas Gross, US Army 97th Field Artillery Battalion, Americal Division. While we were still back on New Caledonia at this time, the food we had was regular Army chow for the most part. I do remember one time, though, that we invited some New Zealand troops up into our area. They brought along some beer and we cooked up half a steer. We showed them how we played polo with our mules using brooms and a soccer ball instead of a polo stick and ball. We all had a great time.

Robert Michel, PhM3/c USN Malaria Control Unit Cactus. One of the island's treasures I bypassed, was the Japanese sake. There were piles of captured crates lying around and some of them were filled with sake. The only problem was, we didn't know if it was good to drink or if the crates were booby-trapped. So, we just left it alone. Looking back on it now, I wish someone would have taken a bottle and tried it. If he had lived, we would have had one big party.

I do, however, remember at one time there was an abundant supply of raisins. We would get a gas can, clean it out well and fill it with water. Next, we would add a 5 pound can of raisins and add some sugar and some yeast that we scrounged from the cooks. In a couple of days, we had raisin wine. God, it was awful, but we drank it, warm and in tin cups.

Marine Sergeant, Section Leader 75mm Pack Howitzers, I/3/11. The torpedoes the Navy used had a propellant that was part alcohol. Guys would get a hold of this and drink the stuff. Guys used to call it torpedo juice. A lot of them would go blind, or it would kill them when they drank it. The problem was, we were on Guadalcanal so long there was a lot of stuff going on.

Robert Michel, PhM3/c, USN Malaria Control Unit Cactus. Scrounging became a way of life for us. Again, we were not really attached to anyone, so we as individuals did our own scrounging and generally shared with our buddies when we made a find of any value. This especially included clothes. Of course, we were always on the look out for a good mess to eat at, especially with the SeaBees.

John Francis Richter, PhM3/c Corpsman USN, 75mm Gun Half-Track Headquarters & Service Battery, 11th Marines. Between our own meager food supplies and that of the more "exotic" food supply left by the fleeing Japanese, plus daily gorging while camped in the coconut grove on an abundance of coconut meat and milk, we survived.

We all attempted to outdo one another in relating what luscious mouth watering menus we would order upon our return to "civilization." Then, in a return to reality, we would take another bite of hardtack!

Here's a couple of factual anecdotes regarding rationing.

On the day the Japanese bombed our half-track position, laying much waste around us, my barnacle laden canteen cup suffered a critical wound. A piece of shrapnel had penetrated one side leaving a gaping hole the size of a dime midway from the bottom. With no replacement available, I found myself short changed in the tasty "beverage of the day" at the field galley. Positively NO liquid could make it past the halfway mark in the cup. There being in place a strict restriction on seconds during our critical short ration period, I sought, and was granted, special "dispensation" from the Mess Sergeant to return with my critically disabled cup for a controversial, and much frowned upon, refill.

Here's another one I call "A Sadist Among Us."

As a Navy Corpsman assigned to a crew of U.S. Marines, victims of a prolonged, severely reduced period of rationing, I found myself desperate to supplement, in some way, our sparse, dull diet. One day, a member of our crew was assigned to a truck detail transporting food from the supply depot to outlying units. After an extended debate, I finally persuaded him to toss something extra tasty and stomach-filling from his truck the next time it passed our bivouac area.

As darkness fell and the truck went by, off flew a large can that came to rest within our perimeter. Setting it aside, we decided to tantalize ourselves by not checking the contents right away. Later that night, we all figured it was time for some goodies. Aided by partial moonlight—blackout was in effect—I was given the honor of reading its contents. Mind you, we were right in the middle of a productive coconut grove.

The contents: shredded coconut! Obviously, we had picked a sadistic, budding comedian for that truck detail.

PFC Stewart Winterton, Marine Rifleman, B/1/7. We had no food! We got used to eating two handfuls of rice per day, the rice we captured from the Japanese. Later on I remember eating Spam, once we got control of the sea and more rations could get in. I only just started eating rice again. I couldn't eat it for years. I got sick of it.

Tom Croft, BM3/c, Gun Loader Turret 1, USS Monssen DD436. The food we had aboard ship wasn't bad. It was always warm, but I don't eat Boston baked beans to this day. We ate them for Sunday morning breakfast and all week long, everyday.

Marine Sergeant, Section Leader 75mm Pack Howitzers, I/3/11. When we landed, we only got in about one third of our food, and not that much more of our ammunition, because the Navy left the next afternoon. If it hadn't been for the Japanese supplies we captured when we first landed, we would have starved to death. I remember the barley and the rice we captured from them. Food was always a problem. We used to catch the iguanas and eat them. They were a little rubbery, but they were OK.

Lt. Cmdr. Lewis P. May, USNR Ret., Assistant Supply Officer, USS President Hayes APA-20. We had excellent storage spaces aboard ship because the Hayes had been built originally to be an around the world liner. We even had a mechanical cow which produced drinkable milk from powder. We got excellent provisions from the Lurline, and I think, the Matsonia, and other ships that were at Espiritu Santo. The Marine and Army passengers loved our chow.

On one occasion at Espiritu Santo, I took a working party on board one of our tank lighters. We went to a provision ship and took on a load of potatoes, which they were anxious to unload so they could depart for the States. We were anxious to have the potatoes because that meant fresh provisions for us. So we loaded the lighter to the gunwales. We proceeded out across the bay. Thank God the water was flat and calm that day as we slowly made our way back to the Hayes and unloaded the potatoes without sinking the tank lighter. That trip is my own personal nightmare and was the closest I ever came to sinking a U.S. Navy boat.

William Ricciardi, MM2/c Boat Gang USS Crescent City. When we hit the Canal in August, we captured everything the Japs had there, including a lot of their food. They had everything camouflaged, but we always found it. In fact, when we started to get prisoners and kept them in a cage on the beach, it was the Jap prisoners who actually showed us how to cook the rice we captured from them. Can you believe it?

Robert Michel, Phm3/c, USN Malaria Control Unit Cactus. When we arrived we had weapons, but we had to scrounge for just about everything else we needed, diesel oil for the sprayers and the dynamite for drainage. Most of the ammo was always laying around for the taking. We had supplies of quinine and Atabrine and all the normal things medics would carry, 1/4 grain morphine syrettes, the sulfa drugs, benzedrine, bandages, etc.

Benzedrine was given out freely to the Marines on the line to keep them awake. I know a lot of us took it at one time or another, but I don't remember the occasion. I do remember, though, the "high strung" feeling it left you with.

In order to understand how and why the military freely dispensed the drug Benzedrine during the war, we should perhaps, first define and discuss adrenaline.

Adrenaline is the body's natural stimulator. It is produced by the adrenal gland located near the kidney. When it is released, adrenaline is carried throughout the body by the bloodstream. Adrenaline directly affects the nervous system and as a result, affects heart rate, dilation of the pupils, and production of saliva and sweat. In other words, it is a stimulant.

Benzedrine is actually a trade name for amphetamines, which are nothing more than synthetic drugs, very closely resembling the body's natural adrenaline. Amphetamines, or as Bob referred to them, benzedrine, when ingested, cause biological responses in the body very similar to that of a rush of adrenaline. Benzedrine is nothing more than a synthetic stimulant causing heightened awareness, euphoria, and a sense of well-being for the user.

Benzedrine was originally used in the 1930s to treat patients suffering from asthma. Later, when the war broke out, benzedrine was used by the military as a stimulant to allow soldiers in combat positions to stay awake and especially alert for long periods of time.

Faced with continuous night actions from Japanese planes, ships, and ground forces, benzedrine was a welcome addition to many Marines' diets.

Robert Michel PhM3/c, USN Malaria Control Unit Cactus. I only administered morphine once, and that was to the entomologist Knight, who was in command of our unit. He had a tent right next to Dave and I.

When it rained, which was often, the coral road right in front of our tent would become very slippery. After one night of strong rain, a flatbed truck, captured from the Japs, came down the road carrying a load of Jap land mines that had been cleared from some area around the Matanikau River. Going up the road in the opposite direction was a jeep. One of the vehicles skidded on the slippery road and the two vehicles collided head on, right in front of our tents.

I was standing at the center pole of my tent shaving when all of a sudden a sort of haze or dust seemed to jump about three feet from the ground. I instinctively dropped to the ground. Then there was an explosion, and where the truck and the jeep had been, there was now nothing but a huge crater in the road. It was a tremendous explosion. I think someone said there were a couple hundred land mines on that truck, and Lord knows what else.

Shrapnel, pieces of the vehicles, and coral flew everywhere. Stuff flew all over the place, including a piece of someone's rib cage that landed right in front of the door of our tent. A piece of shrapnel caught Dr. Knight right in the groin as he was standing in his tent. You would have thought he had been killed.

Anyway, it luckily turned out to be more surprise and terror on his part than anything else. Some pain and a little blood, that's all. But, he was in pain, so I gave him a shot of morphine. At the time, with everything else that was going on around us, it was just considered a scratch, so he wasn't removed from duty. Put on a band-aid and return to duty, that's about what happened. I don't even think he went to the hospital. Usually, the only time one saw the inside of the hospital was when there was some serious injury or sickness.

I often wondered in later years if he had put in for a Purple Heart. Probably not. It was, after all, friendly fire.

...and Then There Were The Critters

Though the South Pacific is known for mosquitoes the size of dive bombers, they also teem with various other kinds of creeping, crawling insect life—spiders, scorpions, ants that could rip through the skin of the toughest old Marine, and centipedes. There was no place on the island that was safe from the invasion of centipedes. A bite from one of Guadalcanal's centipedes could prove not only uncomfortable, but life threatening as well. Many of the centipedes on the island were not only a frightening nuisance but were poisonous to boot. A bite from a centipede at the very least caused violent vomiting that could last for hours. Reaching a length of ten to twelve inches, the centipede could strike a victim from either end of its body. Native islanders used to say the first day after being bitten by a centipede, you thought you were going to die; the second day after being bitten, you were afraid you wouldn't.

Along with its abundant insect life, Guadalcanal presented another side to an idyllic life in a South Pacific tropical paradise, complete with fifteen foot crocodiles, thousands of rats, bats, and surrounded by shark-infested waters.

Robert Michel PhM3/c, USN Malaria Control Unit Cactus. On the Canal once, I saw an alligator, or maybe it was a croc. It was at a stream or small river the Marines gave the name Alligator Creek. I don't know. I did see a lot of iguanas and for target practice once shot a big green one. Got him on the first shot. Another time, in an area that was fairly well cleared out except for some lime trees, I saw a green snake up in a tree that must have been four or five feet long. I left him alone.

I sure do remember the centipedes, though. One night while lying on my back on my cot, I saw this huge centipede crawl up to the top of the outside of the tent, then crawl inside through the hole at the top of the tent pole. He started to come down right over my mosquito bar, and just dropped. The damn thing couldn't have been more than a few inches from my face. You talk about fear. We were able to knock him off onto the floor and kill it.

Marine Sergeant, Section Leader 75mm Pack Howitzers, I/3/11. I never saw a snake the whole time I was there. Lots of centipedes and other things. Crocodiles, they were all over the place, especially in the Lunga River and the Ilu, which you hear called the Tenaru.

Every night we would put out what we called listening posts in front of our lines, because the Japs were always sneaking up on our positions or trying to infiltrate our lines. A listening post was usually just a couple of guys we would send out in front of our positions about fifty to a hundred yards who could warn us of any Japs moving around.

One night we had sent these two guys out and everything was pitch black and very quiet. The guys started hearing a scratching noise in the underbrush. There would be silence, then the scratching noise again. The scratching kept getting closer and closer to one of the guys in his hole. He waited until the scratching noise was right at his feet and he opens up with five shots. All he heard after shooting was a thrashing noise in the brush. He then gets up and starts hauling ass back to our section's position.

The noise of the firing alerted everyone back on the line. Everybody was wondering what the shooting was all about, and then they heard this guy screaming, "Japs. Japs. They killed my buddy." The guy was running back to our lines as he was screaming at the top of his lungs. Everyone on the line was scared to death, just waiting for an attack by the Japs. We were all real tense, wondering what would happen next. Suddenly we hear this guy's buddy heading back into the lines. It was a good thing he remembered the password or he might have gotten shot. The next thing you know he had found his buddy who had done all the shooting and started to beat the shit out of him right there in our positions. All he said was, "You killed an iguana, you son of a bitch."

Robert Michel, PhM3/c, USN Malaria Control Unit Cactus. My first vivid experience with ants was up around the Matanikau, seeing them go in and out of the eye sockets of dead Japs carrying pieces of human tissue. Also prevalent, though, were those real little ants that we called piss ants. One day while walking in the bush, I brushed up against a bush about seven or eight feet high. It looked like any other bush, except that the entire thing was covered by millions of piss ants. When I brushed against it, the ants all dropped on me and were caught in the beads of sweat all over my body. The boys with me immediately ran their hands up and down my body to kill them. Not before I was bitten pretty good, I might add.

Souvenirs

Robert Michel, PhM3/c USN Malaria Control Unit Cactus. Souvenirs, too, were fairly valuable to those guys who had access to the crews of the ships. They traded or sold their souvenirs to the ships' crews for whiskey if they could get it. We, however, didn't have access to the ships or crews because of our location on the island.

I remember I had a Jap rifle, a Jap bugle that I had picked up, and a knife. I also fashioned an armband out of a piece of aluminum from a downed Jap Zero. I hammered it out flat and bent it to fit around my upper arm. It had some inscription on it. One of the SeaBees actually made my knife out of a file. He heated the file red hot to take out the temper and then fashioned it into a two edged blade, and then re-tempered it. The handle of the knife was from the plastic handle of a shaving brush and pieces of aluminum from a Jap Zero. It was a beautiful knife that I brought home with me. The rifle I gave to someone while I was still in the Pacific, same with the bugle. The knife I finally threw away just a few years ago.

Marine Sergeant, Section Leader 75mm Pack Howitzers, I/3/11. I had all that stuff, Jap swords, flags, everything. It was about 1963, and I went down in the basement and looked at all those souvenirs hanging up and thought to myself, What am I keeping all this junk for? You know what I did? You're right, I threw it all out. The only thing I kept was a Jap coin that to this day still has the dried blood of a dead Jap on it. I gave that to my son and said, "Here, it's yours to keep."

William Ricciardi, MM2/c Boat Gang USS Crescent City. I still have a Jap helmet and a Jap sword that's a little different from most of the Samurai swords you've seen. It looks like a Samurai sword except the blade is only about two feet long. Attached to the scabbard for the sword is another scabbard with a shorter sword with a blade about a foot long. I think this was the type of sword the Japs used to commit Hari Kiri with. I still have the Bowie knife I carried with me on the Canal. The only thing wrong with it is the very tip of it is bent. The tip bent when I stuck it in a Jap and it went all the way in and hit his spine.

Robert Michel, PhM3/c USN Malaria Control Unit Cactus. When I first arrived on the Canal, I ran into some Marines who claimed to be China Marines. Anyway, these older guys had earrings in the left ear and told me it was because they served in China before the war. I remember I almost did one myself. Both Navy and Marine guys on the Canal started wearing the same kind of earrings in their left ear as a symbol of their time on the Canal. I did end up piercing some ears on other guys, at their request. I stuck a threaded needle through the ear, put some merthiolate on the wound, and told the guy to keep pulling the thread back and forth to keep the hole open. No charge.

William Ricciardi, MM2/c Boat Gang USS Crescent City. A lot of guys took to wearing an earring in their left ear to show they were there for the hit on the Canal. Some guys even wore two. I don't have mine anymore—I lost it somewhere along the way. One thing that bugged me was the guys who wore them who were never even there. The Marines were wearing them, and the Navy guys, too.

Dave Minton PhM3/c, USN Malaria Control Unit Cactus. I still have mine. I don't know where I got it. It was round and had one end that was sharp that hooked into the other end to close it. You'd stick it through and it was hinged so it would close together when it was in your ear. It was a beaut. The hole has since closed up and healed over. I can just feel a lump in there now where the hole used to be.

Marine Sergeant, Section Leader 75mm Pack Howitzers, I/3/11. I believe that some guys may have got away with that, but not in our outfit. I was in the Marine Corps for six and a half years and never saw anything like that the whole time. Can you imagine what the top sergeant would have done if he saw an earring in my ear? He would have just reached up and not only ripped the earring out of my ear, he would have ripped my whole ear off and rammed the earring and my ear right up my ass.

Tattoos, now that was another thing. A lot of guys had tattoos; all you had to do was roll your sleeves down and cover it up so no one could see it.

Chapter 7

Guadalcanal: The Campaign

August 1942-October 1942

As August rolled into September, life for the troops stationed on Guadalcanal turned into a monotonous routine. Daily air raids delivered by Japanese fighters and bombers flying out of Rabaul kept the survival-savvy men on the island away from Guadalcanal's number one target, Henderson Field, between the hours of 1130 and 1430 (11:30 A.M. and 2:30 P.M.). Mastery of the surrounding seas changed hands literally every twelve hours. The American Navy ruled the channel from dawn till dusk, offloading desperately needed supplies from transports churning the waters up and down the sound. Once the afternoon shadows started to lengthen off the tops of the island's coconut trees and the sun cast an orange glow over the waters of Lengo Channel signaling the approaching dusk, American ships beat a hasty retreat, heading for safer waters away from Guadalcanal. Once the full darkness of night enveloped the island, the Japanese Tokyo Express cruised into position off the island. Japanese troopships and their escorts quickly dashed into shore unloading fresh troops and supplies for their beleaguered forces. On their way out, Japanese ships never failed to lob a few well placed rounds into U.S. shore positions.

By the first week of September, the Tokyo Express had strengthened the Japanese 35th Infantry Regiment under the command of Major General Kiyotake Kawagushi to over 5,000 men who were massed in the vicinity of Taivu Point, just sixteen miles east of the Marine perimeter. These nightly forays by the Japanese resupply convoys, many consisting of not much more than motor powered barges laden with men and supplies and escorted by tiny destroyers, were soon referred to by the Marines on the Canal as “rat transports.”

As a result of lack of food, very little rest from the daily air raids and nightly naval shelling, the Marines, Navy shore personnel, and air crews were beginning to show signs of fatigue. Scrounging for food had become a way of life with daily rations in constant short supply and of very poor quality. Day-to-day existence had turned into an ordeal that Vandegrift had to take into consideration as he continued his efforts to locate the Japanese and extend the limited confines of his perimeter.

From the first day of the landing on August 7 until the first week of September, activity for the men on the Canal continued unabated at a fevered pitch. The Marines had by now handed the Japanese their first resounding defeat in battle, stacking up almost 800 enemy bodies on the banks of Alligator Creek on the eastern edge of the Marine perimeter. (To the west, the Navy once more met a superior Japanese Naval force but was able to turn back Japanese efforts to retake the airfield on Guadalcanal. During the Battle of the Eastern Solomons and the ensuing melee of combat, the U.S. Navy was dealt a crippling blow when they lost the carrier Enterprise to Japanese naval bombs.)

The burgeoning Cactus Air Force continued to grow in size and strength, receiving reinforcements from the Army Air Corps in the way of a squadron of P400s. Orphaned Naval squadrons from crippled or destroyed carriers also flew in. They had provided air cover over the waters surrounding the island.

As air traffic into and out of Henderson Field progressively escalated and Japanese air and naval bombardments continued unabated, an additional airfield was desperately needed, if for no other reason than to protect valuable aircraft from possible destruction on the ground at the Jap's number one target, Henderson Field.

Fighter One

On September 9, Fighter One airfield opened for business on the kunai grass plains just one mile east of Henderson Field. Built on the flat, hard ground that lay just beneath the burnt off tropical grass, Fighter One would soon provide the Cactus Air Force the safe space required for dispersal of their aircraft during the coming weeks of battle.

Marine Sergeant, Section Leader 75mm Pack Howitzers I/ 3/11. We decided to solve the problem of the Japs sneaking around in front of our positions by burning off the kunai grass. We burned an area about 300 yards wide in front of us. Underneath that kunai grass the ground was beautiful; it was flat, solid ground. The area was perfect for airplanes and that's where they ended up building Fighter Strip One.

I remember the first planes that ever flew into Fighter One. They were F4F Wildcats. The first two came in and the pilots hopped out onto the ground. The third plane I'll never forget. The pilot came in for a perfect landing on that new grass strip, the only thing was he forgot to lower his landing gear. Just a mistake in his landing checklist, I guess, but anyway that plane came in and tore itself up as it skidded across the runway. When it finally came to a stop, out jumps the pilot, a big red headed guy who had a big

NAVY DEPARTMENT COMMUNIQUÉ

No. 124, September 15, 1942
South Pacific (all dates given are east longitude)
Based on reports received up to 6 P.M. (e.w.t.).

1. Japanese attempts to recapture U.S. positions on Guadalcanal Island have been intensified. Heavy fighting has been in progress since the night of September 12-13 between our Marines and reinforced enemy troops on the island. Details of the fighting are lacking, but reports received to date indicate that the Marines are maintaining their positions.

2. During the past few days the Japanese have increased the intensity of their bombing attacks by aircraft and the bombardment by surface vessels. Some attempts by the enemy to land small detachments of troops have been successful.

3. On September 11 Army Flying Fortresses shot down four Zero fighters in the vicinity of Guadalcanal and strafed shore installations on the island of Gizo. On September 12 Army bombers shot down two enemy seaplanes.

4. During the night of September 12-13 our positions on Guadalcanal were shelled by enemy surface craft. One of these vessels was reported to have been hit by our shore batteries. Contacts with enemy patrols were made during this same night and heavy fighting continued during the following day and night.

5. On September 13 two flights of about 28 enemy bombers each, escorted by Zero fighters, attacked the airfield at Guadalcanal. Navy and Marine Corps Grumman Wildcat fighters shot down 4 bombers and 4 fighters. This same day Army bombers attacked seaplanes on the water of Rekata

Bay on the northwest shore of Santa Isabel Island. One enemy plane was destroyed and a second was damaged.

6. During the night of September 13-14 our positions at Guadalcanal were shelled by enemy surface craft and our troops engaged in heavy fighting with enemy troops, who made an unsuccessful attempt to capture the airfield.

7. On September 14 an enemy flight of 28 bombers escorted by fighters, attacked Guadalcanal shortly after noon. Navy and Marine fighters shot down 1 bomber and 5 fighters.

smile on his face. I'm sure he was happy to survive, but I'm also sure he got chewed out pretty good for destroying an airplane. At that time planes were a lot more scarce than pilots were.

All the guys who were in positions around Fighter One used to stand up and cheer when our planes would come in to land there. The pilots would fly over real low and do a roll before they landed. One roll for each Jap plane they had shot down. One roll, one Jap; two rolls, two Jap planes, and so on. You could hear the cheers from the guys as the pilots would do those rolls. We were real happy to see how many Jap planes were being destroyed by our guys.

Enduring the chaos of battle, constant enemy bombardment, and continued lack of food and supplies, the men of the 1st Marine Division daily continued to carry out their assignments. From front line riflemen to cooks and bakers, armorers and mechanics, pilots, radio operators, medical personnel, stevedores, and construction workers, each contributed in their own way to the enduring presence of the Marine garrison. Malaria Control Unit Cactus was no different. Day in and day out the Skeeter Beaters, an integral part of the Marine team, loaded up their backpack sprayers and pursued the *Anopheles* into her breeding grounds; the maze of swamps, lagoons, water filled ruts, and creeks that blanketed the tiny perimeter encircling Henderson Field.

Interestingly enough, it wasn't until mid-September or early October, when resupply efforts became more consistent, that the men of MCU Cactus finally received a welcome addition to their meager arsenal of equipment.

R & D: Developing Products for a Better World

Robert Michel, PhM3/c USN Malaria Control Unit Cactus. One day in mid-September, or maybe it was early October, a couple of wooden crates were delivered to our compound. Somebody Stateside had sent us boxes of insecticide bombs to evaluate and experiment with. They were heavy metal canisters about five or six inches tall and about four inches in diameter. They resembled an oversized hand grenade in shape. When you pressed the valve on top, it would spray insecticide that killed adult mosquitoes. We used these primarily to spray inside the mosquito netting in our tents before getting into bed.

We only received about twenty-four canisters, and I have no idea if any paperwork accompanied them or not. I don't remember any markings, although I am sure there must have been instructions with them, otherwise we wouldn't have known what they were for. We never received any more canisters after that first shipment.

They were, I believe, the first aerosol insecticide spray cans ever tested in the field, so to speak. We called them a freon bomb, for lack of a better name. The label must have mentioned freon because I'm sure we had never heard the word before and didn't know what it meant. We loved them and told the brass they were great. Now look at what's happened with aerosol sprayers since then.

Dave Minton PhM3/c, USN Malaria Control Unit Cactus. Oh yeah, I remember when we got those things. If I recall, they were round, about five inches long, and about four or five inches around the middle and made of steel. They were heavy, too—about as heavy as a hand grenade. If I remember it right, they were round at the top and the bottom. They wouldn't stand up by themselves because they had that round bottom.

John Francis Richter, PhM3/c Corpsman USN, 75mm Gun Half-Track, Headquarters & Service Battery, 11th Marines. Regarding the aerosol spray insecticide containers, I recall their solid, end-rounded metallic look. I honestly cannot recall ever using them on Guadalcanal in our isolated half-track positions. I believe I personally came in contact with them at a later date while in Australia and quite possibly in Cape Gloucester, New Britain. I can't recall any details regarding their effectiveness, however.

Arthur King, MD U.S. Army Surgeon, Medical Supply Officer, South Pacific. There was a guy back in the States who was building a lot of these experimental gadgets and sending them down there for these guys to test. I didn't see the bug bombs myself when I got there, but I knew of them and knew they were experimenting with them. The SeaBees used to love that kind of stuff. They were always tinkering around with something—sometimes acting like they didn't have anything else to do.

As amazing as it sounds from Bob's description of the aerosol insecticide canisters MCU Cactus was provided, they had to be some of the first ones ever used by U.S. troops overseas in World War II.

The concept of pressurized dispensers, though, was not really a new one. Pressurized, carbonated beverages were first introduced in France 150 years earlier in 1790. A patent for aerosols pressurized with ethyl and methyl chlorides used as propellants was issued in the United States in 1899. Aerosol fire extinguishers were developed in 1933 by General Motors and were sold with their automobiles as an accessory. Little was done in the way of research and development of aerosol insecticides, however, until the start of World War II when their need became glaringly apparent as war in the Pacific loomed.

With insect propagated diseases such as malaria and dengue fever striking down hundreds and later thousands of healthy U.S. servicemen, government officials in the Department of Agriculture gave top priority to the development of an effective, hand held bug blaster. Two men within the department, L. D. Goodhue and W. N. Sullivan, came up with the first workable design that they patented on October 3, 1941, right about the time Bob Michel was well into his medical training at Brooklyn Naval Hospital.

Constructed of two steel halves welded together to form the canister, and with the weight of a hand grenade, the bug bomb (as it was aptly nicknamed) was ready to go. Pressurized with a liquefied gas—actually fluorocarbons that we hear so much about today—the bug bomb became the precursor for everything we now take for granted that is dispensed from a spray can, such as shaving cream, deodorants, hair sprays, lubricants, and cleaners. As Bob puts it, "Now look what's happened with aerosol sprayers since." Little did he know at the time that he would be providing valuable market research and product development services for a future industry that would become worldwide.

As its effectiveness was proven by MCU Cactus, it wasn't long after the Guadalcanal campaign that U.S. troops around the world were spraying the bug bomb at anything that crawled, flew, or hopped within their reach.

September 18, 1942

At dawn on the morning of September 18, 1942, two U.S. destroyers, Monssen and MacDonaugh, slipped into Lengo Channel just west of Lunga Point, screening six Navy transports carrying the men and equipment of the 7th Marines. Three battalions totaling almost 1,800 fighting men were on board. Nicknamed the First to Fight, the 7th Marines had been dispatched to Samoa the previous February, one of the earliest Marine Defense Forces deployed in the Pacific. Now their months of training in the tropics would serve them well in the hard months ahead.

Circling the transports like mother hens, USS Monssen and MacDonaugh opened fire with their five batteries to provide covering fire on suspected Japanese positions near the Matanikau River, while the men and equipment were hustled over the sides of the transports and rushed ashore. The cargo ship Bellatrix dashed in close to shore, offloading her cargo of aviation fuel and barbed wire before the afternoon's enemy air raid on Henderson Field started. The arrival of the aviation fuel and barbed wire were the first such supplies delivered since the Marines landed the month before.

PFC Stewart Winterton, Marine Rifleman, B/1/7. One of the things I remember most about that first afternoon coming ashore is that a Navy PBY was shot down by our own ships. He came in over the channel, out of the sun the wrong way, violating all rules of approach. They all opened up on him.

As the men of the 7th stepped off the Higgins boats, they were immediately put to work gathering their equipment and humping it off the beach to a position in a coconut grove about 500 yards inland. It was here in the coconut grove that their officers ordered the men into their first bivouac on the island.

PFC Stewart Winterton, Marine Rifleman, B/1/7. My first impression of Guadalcanal was, What in the hell do we want this place for? It's nothing but jungle. All the coconut trees we saw were blown away by shells.

We were shocked when we saw the Marines who had been on the island for the past month. As it turned out, though, in another month, we looked the same. I guess you could say they were glad to see us, though all they would tell us was, "You'll see what it's like," or, "You'll find out."

I was just twenty years old when I landed on Guadalcanal. I wanted to survive. Others were as young as seventeen and cocky—not me. Mostly we were just scared.

Oh yes—the Japanese ruled the sea then. Our ships just dropped us off and took off. There was a constant cat and mouse game going on between the Japanese and the U.S. Navy around there.

Exhausted, filthy Marines who had been on the island for over a month now were milling around and scrounging what they could from the new arrivals. Having been aboard ship for over two weeks, the reinforcements of the 7th collapsed, throwing their equipment to the ground. Officers and noncoms ordered the men to disperse and dig in. Most disregarded the orders, just glad to relax on ground that wasn't rolling or pitching—thinking there would be plenty of time to dig in tomorrow after they had gotten a good night's sleep.

Out in the channel, the transports and destroyers fired up their boilers and beat a hasty retreat away from the island, much the same way the first Task Force departed after the initial Marine landings back in August.

Task Force 65 and the Marines of the 7th represented the first reinforcements, as well as the first real load of supplies, to be brought to the island in the last month. For the first time since their arrival, veterans of the past month's fighting could count on getting more than just two lousy meals a day.

Vandegrift was now ready to expand his perimeter to the east beyond the Ilu River and west of the Matanikau River. With fresh troops at hand, the General could implement his plan for an offensive defense stretching the confines of his perimeter and placing more distance between Henderson Field and Japanese artillery. The arrival of the 7th finally gave Vandegrift the opportunity to once and for all move out and locate his opponent.

Robert Michel PhM3/c, USN Malaria Control Unit Cactus. It was right about this time that I was back in a secondary area, not on the line, and there were a lot of Marine replacements who had just gotten there. I don't really remember why we were there, but I do remember I got into an altercation with one of those new guys over something or other. I probably called him a name or something. Anyway, this guy started to beat the hell out of me. I'm a pretty big guy, over six feet tall, but he was even bigger. He got me down on the ground and was on top of me.

Wouldn't you know, this Marine Gunny Sergeant, who was considerably older than the both of us and looked like a wrestler, reaches down and grabs this guy by the back of his utilities. With one arm, he lifts the guy up, spins him around, and punches him right in the face, knocking him back to the ground. The Gunny stood over that guy, looking down at him and told him in no uncertain terms, "Don't ever fool around with the Doc. Someday you may need him."

I guess by this time, all of us were starting to get a little edgy from the strain, and this kind of stuff was probably the result of nerves.

From this day forward until their rotation to Australia in December and January, U.S. Marines fought and annihilated the best the Japanese could throw at them. Once again, it is not the intent of this book to rehash in detail every trial or act of heroism that took place the first four months on Guadalcanal, but merely to offer a sense of, or general empathy for, the veterans' day-to-day life on the island during these trying times. However, in an effort to provide context for the reader, the two-week period from

September 19 to October 1, 1942, will always be remembered for the battles that took place on the banks of the Matanikau River and the attempted Marine encirclement of Japanese positions west of Point Cruz, known today as the 2nd Matanikau.

Japanese Landing Craft
Guadalcanal, 1942

PFC Stewart Winteron, Marine Rifleman, B/1/7. The problem I have today is that we went on so many marches and had gotten into so many skirmishes that I cannot really remember one from the other. There were so many ambushes that they are hard to distinguish—all those actions blur into each other.

October and Here Comes the Army

By the first of October, General Vandegrift realized he was fighting not only the Japanese for control of the island, but also the inimitable factor of time. The longer his troops remained on the island with no relief, the more physically drained they became, dropping every day now from the effects of malnutrition and disease. A quick glance through the daily casualty reports indicates that most devastating among the many maladies the Marines suffered was malaria.

Vandegrift, recently reinforced and with adequate supplies of ammunition, continued to pester the Japanese in an effort to expand his perimeter outward and to place more space between Japanese artillery and his precious airfield. Japanese military commanders, however, were far from ready to concede defeat and give up the island to the Marines. In fact, they had finally decided the time had come once and for all to knock the upstart Marines off the island and right back into the sea from whence they came. Japanese plans for just such an offensive were rapidly being put into play calling for the immediate destruction of their number one target, the airfield. Once Henderson Field was rendered inoperable, the Japanese thought they would be free to land in force unopposed, making short work of the American defenders at a time and place of their choosing.

Throughout September and into early October while the Marines continued to hang onto their tiny piece of real estate on the Canal, Admiral Ghormley (Commander South Pacific) and General Harmon (Commander Army Forces South Pacific) had been pressuring the Joint Chiefs of Staff for additional troops with which to reinforce Vandegrift. Harmon and Vandegrift both agreed with Ghormley's assessment that unless reinforcements were sent to the island soon, the Japanese could very well threaten the security of the perimeter, if not retake the island itself within a short period of time.

On the morning of October 9, 1942, the 164th Regiment, a North Dakota National Guard unit assigned to the Americal Division, loaded up its men and equipment and headed out for the island of Ndemi. Their mission was to secure the island for construction of a new airfield. Shortly after their departure, the constant pressure from Ghormley, Harmon, and Vandegrift paid off when the troopships were abruptly rerouted and dispatched to Guadalcanal, where they arrived just four days later off Lunga Point on the morning of October 13.

Shortly after the Army's 164th Regiment landed on the beach, Japanese bombers swept in over the island at 1130 hours dropping their bombs on Henderson Field. The Japanese raid was successful in hitting the runway, storage areas filled with drums of aviation gas, and a number of Cactus Air Force planes parked around the airfield. No sooner had the first wave of planes turned back toward their bases on Rabaul than a second wave of enemy planes hit the airfield, this time exclusively targeting the runway, which sustained nominal damage. The 2,850 men of the 164th indeed received a warm welcome by the Japanese, one that was only to get warmer as the day continued.

At 1830 hours that afternoon (6:30 P.M.), Japanese artillery started lobbing their heavy shells into the western end of Henderson field. With the Cactus Air Force grounded from the morning's activities and the Japanese artillery positions just outside the range of Marine artillery emplacements, there was nothing the Americans could do but sit back and take the pounding the enemy was dishing out.

While Japanese artillery was raining down on Marine and Army positions all over the western end of the Lunga perimeter, two Japanese battleships, the Kongo and the Haruna, were steaming down the Slot in hopes of adding the coup de grace to the already heavily damaged Henderson Field. Sailing unopposed into their final firing positions just offshore, the Kongo and Haruna each mounted eight, fourteen inch guns that, when armed with high explosive shells, would deliver more terror and damage than anything that had hit the island yet that day.

At 0130 hours the morning of October 14, the Japanese battleships opened fire. Within the first hour of their shelling, the two battleships delivered over 100 high explosive projectiles into Henderson Field. Riding out the storm in all manner of defensive positions—from coconut log covered bunkers to slit trenches—the Americans, once again, were unable to fight back. Almost immediately shells ripped into the aviation fuel dump, sending

NAVY DEPARTMENT COMMUNIQUÉ
#152, October 14, 1942
South Pacific

The following communiqué is based on reports received up to 7:30 P.M.., October 14 (Washington Time): (all dates are east longitude).

1. On the night of October 13-14, Japanese surface vessels bombarded our airfield and shore installations on Guadalcanal.

2. Information has just been received that during the early morning of October 15th, troops from enemy transports covered by naval units were landing on the north coast of Guadalcanal to the westward of our positions.

flames hundreds of feet into the air and destroying what little fuel reserves Cactus Air Force had left. As the burning fuel lit up the surrounding night, grounded airplanes were destroyed by the continued shelling. By morning it was clear to the defenders they were now without an air force. In all, the Kongo and the Haruna lobbed a total of 973 fourteen shells into the airfield and its immediate surrounds while Japanese bombers continued to circle around the edges of the perimeter dropping bombs on the Marine positions for the rest of the night.

By morning, with no end to the hellacious beating the Americans were taking in sight, at 0530 hours Pistol Pete started dropping high explosive artillery shells randomly throughout the area.

To date, this had been the most devastating and lengthy bombardment the U.S. troops had yet endured. As the smoke and dust finally started to clear, it was determined that out of a total of over 23,000 U.S. servicemen on the island, miraculously, only 41 had been killed.

Arthur King, MD, US Army Surgeon, Medical Supply Officer, South Pacific. I went over to Guadalcanal in early October right around the time the big Japanese naval bombardment started hitting there. My job was to find out what the Americal Division would be up against when they began supporting the Marines.

William C. Dickman, MSgt. USMC Enlisted Navigator/Naval Aviation Observer VMJ-253 MAG-25, South Pacific Combat Air Transport, SCAT. I was on the Canal on October 14th and had to stay overnight there because of the huge Japanese naval bombardment. They really worked us over during that battle in October. They set some of our planes on fire. I slept in a foxhole that night due to the close shelling. Luckily they didn't hit our transport plane.

There was another good size battle shortly after this shelling at the end of October. We were flying gas, bombs, ammo, and food into the Canal and flying wounded guys out to Efate and Espiritu. We were flying missions around the clock because we were so short handed and our other squadrons were not to arrive until November. On October 30, 1942, on one of our flights into the Canal, I was wounded, but luckily not severely.

Marine Sergeant, Section Leader 75mm Pack Howitzers, I/3/11. For three days straight, October 13, 14 and 15, Jap ships shelled us nonstop. That was the day the Army arrived. Those guys got bombed and shelled all day and through that whole first night. That was really something else. There was nothing you could do except loosen the chin strap on your helmet and get in your hole. We dug two-man and three-man slit trenches and would just sit it out. The Japs were firing armor piercing shells, and unless they landed right on top of you, they wouldn't kill you. They could hit a few feet away and you would be OK. They would make a hell of a hole in the ground, though, and they were really loud.

The shells were going right over our heads and landing about a hundred and fifty to three hundred yards from us. They were hitting around the airport there and killed a lot of pilots and destroyed a lot of the planes on the ground.

Robert Michel, PhM3/c USN Malaria Control Unit Cactus. The October Jap Naval bombardment of Henderson Field is one which I will long remember. Our tents were almost in a direct line with the Jap ships in the channel and the airfield. As a result, every shell that came over came right over our compound. There is nothing more terrifying than the sound of shells passing over your head. They all hit fairly close to us as well.

I sure do remember the Army reinforcements who arrived in October. I believe it was the Americal Division. We scrounged some new clothing off them.

PFC Stewart Winterton, Marine Rifleman, B/1/7. When the Army came in, many of us Marines started showing up in clean Army uniforms.

William Ricciardi, MM2/c Boat Gang USS Crescent City. When the Army got there, they brought in plenty of booze for us to steal. They brought in so much stuff we were asking them where the USO show was. Can you believe this? They even put MPs on the beach. They actually brought them with them. The Marines never did that. Another thing was they started this saluting thing. Their officers expected to be saluted just like they were in the States. We told them we don't salute anybody on the Canal. There were always Jap snipers around and saluting just let the snipers identify our officers. They liked to target them anyway.

We would go back and forth to Noumea all the time to pick up supplies and bring them back to the Canal. A group of Army officers over there, when they learned they were shipping out for the Canal, decided to hide a load of whiskey in a pontoon. They talked a

NAVY DEPARTMENT COMMUNIQUÉ

#153, October 15, 1942
South Pacific
(all dates are east longitude).

Dispatches from our forces in the South Pacific reveal the following chronological developments leading up to the current battle in the Guadalcanal area.

On October 13:

1. During the afternoon, the airfield at Guadalcanal was twice bombed by enemy aircraft. Three enemy planes were shot down and one U.S. fighter was lost.
2. By nightfall U.S. auxiliaries had landed reinforcements for our troops at Guadalcanal. Although these ships were attacked by enemy bombers, no damage was suffered and our ships unloaded and withdrew.
3. During the night of October 13-14, the airfield and shore installations at Guadalcanal were heavily bombarded by an enemy surface force believed to have contained battleships, cruisers, and destroyers (reported in Navy Communiqué No. 152). Shore batteries scored three hits on enemy destroyers during the bombardment.

U.S. Army troops are participating in the defense of Guadalcanal.

Navy Ship's Fitter into cutting a hole in the pontoon and they stashed their whiskey in there. They had the Ship's Fitter weld the hole in the pontoon back up again. They marked the pontoon with X69, or something like that, so they would know which one had the whiskey in it when the cargo was unloaded on the other end. We took on the whole load of pontoons on that trip and had them stored in the number three hold of the ship. When we got back to the Canal, they had one officer in the hold and one on the deck to keep an eye on the cargo as it was brought up to the deck in the cargo net. We had already stolen the pontoon with the whiskey in it, though. The Army officers were really upset when they found out it was gone.

Robert Michel, PhM3/c USN Malaria Control Unit Cactus. The word was out that when the Army relieved the Marines up front it was only to lose what the Marines had already gained. The Marines had to go back up and do the job all over again. At least that's the story that was going around at the time. Who knows if it is really true or not? I do know that at the time the Army was not liked very much by the Marines.

Marine Sergeant, Section Leader 75mm Pack Howitzers, I/3/11. When the Army came in on October 13, they were a great new source of supply for us. They carried everything with them in what we called sea bags—they called them duffel bags. We couldn't believe all the stuff they were carrying. They were big, raw-boned boys too, from the Dakotas. They made us look puny. They were the 164th Regiment, National Guard.

They brought the M1 Garand rifle with them, .30 caliber, semi-automatic, gas operated, and eight shots. The Garand would change the whole concept of our firepower for the Japs. You see, the Japs and us were both armed with five shot, bolt action rifles; we had the Springfield .30 caliber and they had a .25 caliber. We used to get in duels with the Japanese and would count the shots they fired, and they would count ours. We would count their shots until we heard five, then try to reload fast and get the drop on them while they were reloading. The story goes that a Jap and one of our Army gyrenes started shooting. When the Jap had counted five shots, he stood up with his rifle aimed, ready to catch our guy reloading. The gyrene saw the Jap stand up and put three right into him with the Garand.

Japanese Light Machine Gun
Guadalcanal 1942

Lt. Cmdr. Lewis P. Mays, USNR Ret., Assistant Supply Officer, USS President Hayes APA-20. On one occasion, we delivered Army troops to Guadal and took aboard Marines whom the Army was relieving and substituting for with their own men. One of the Marines, very badly afflicted with malaria, was a Captain Jim Robinson, who had been my college roommate at the University of Washington. When he was brought aboard, he was allowed to bunk in my quarters on the Hayes. We took him and the rest of the Marines to Brisbane, Australia, where he eventually recuperated. Later, Capt. Robinson participated in some active landings with the Marines at Pelelieu and Iwo Jima, where he was badly shot up and again returned to Australia. It was while he was in Australia that he met and married his wife, Betty. They later had two sons and still live in Brisbane to this day. We have visited back and forth several times since the war.

On one occasion, I think it was a night action, we were enroute to Guadalcanal with Army reinforcements and we were attacked by seven or eight Mitsubishi torpedo bombers. They had been trailing us at dusk and then at nightfall launched their attack on us. Our Admiral had been watching them all this time and he ordered our ships to turn so that the enemy planes came down between our ships and not across our beam. We aboard the Hayes, and those on our other ships, sprayed them with .50 caliber machine guns and shot down seven or eight of them as each one made his torpedo run. In the turmoil of the attack with our ships all firing at the planes, some of our rounds were hitting our own ships.

Quinine, The Coolidge, and Malaria Control

The President Coolidge was one of the world's last truly luxurious ocean liners ever launched when it was christened in 1931. An impressive 654 feet in length, it contained two saltwater swimming pools (one which actually had an artificial sand beach), a fully equipped gymnasium, a children's playroom, and a huge recreation deck for golf, tennis, handball, and squash. The ship boasted a stock exchange for those passengers wishing to follow the market on their voyage, as well as a shopping arcade, a soda fountain, beauty salons and a barbershop. The Coolidge was a popular cruise ship on the Sunshine line, plying its way from San Francisco to Honolulu, Yokohama, Shanghai, Hong Kong, and Manila.

When the war started in 1941, the Coolidge along with many of her sister ships in the old Presidential Line, were drafted into military service and converted to troop transports. When the Coolidge was fully operational and ready to serve her new military masters, she was capable of carrying a total of 5,000 troops and all of their equipment. One of her swimming pools had been converted into a cargo hold. Additional toilets were installed and hundreds of bunk beds added. One notable feature of the ship, however, was not removed. The famous Lady and Unicorn decorative panel above the massive fireplace in the First Class Smoking Lounge remained in place. The Coolidge would earn an infamous place in history after making a number of runs to the South Pacific during just a few short months in 1942.

It was on October 6, 1942, that the President Coolidge departed San Francisco bound for New Caledonia, with her final destination scheduled for Espiritu Santo in the New Hebrides. She carried the 172nd Regiment, 43rd Infantry Division. At Espiritu Santo, the 172nd was scheduled to lend a much needed hand to the beleaguered Marines on Guadalcanal and were to be some of the first reinforcements for the Marine garrison still on the island.

Of much greater importance perhaps than the Coolidge's cargo of reinforcements earmarked to assist the Marines on Guadalcanal, was her cargo of 591 pounds of precious quinine. Unknown to most, the Coolidge had in her holds the entire stock of quinine available to the United States military at that time in the war; an amount equal to over 250,000 single doses of the lifesaving drug. Difficult and expensive to produce, quinine was in incredibly short supply, and it seems the Navy had committed all her quinine eggs into one basket in the hold of the President Coolidge. Once the Japanese had driven the British, Australians, and Dutch out of Java and the Netherlands East Indies, the world's most prolific source of the cinchona tree and its quinine bearing bark, the Allies were left with synthetics as their sole malaria treatment arsenal.

Arthur King, MD, US Army Surgeon, Medical Supply Officer-South Pacific. I begged the Chief Medical Supply Officer for the entire South Pacific not to put all that quinine on one ship. I told him the shipment should be split up amongst a number of ships because it was so valuable. I begged him to go to the G-4 on this, but I was ignored. They all said nothing could happen—it would be safe on the Coolidge. That shipment of quinine represented every last ounce available in the States to us at the time. It was difficult to produce and that was all we had.

Two weeks later, to the day, on October 20th, the Coolidge sailed into Espiritu Santo through Segond Channel. A civilian ship under contract to the military, the Coolidge was captained by a civilian. Ironically, it was that fact, coupled with the intransigence of the Navy, that would soon result in one of the greatest disasters of World War II.

The Navy had littered the waters of Segond Channel over the past few months with a complex pattern of mines and held the map to the mine fields under the strictest security directives. The Captain of the Coolidge, because he was a civilian, lacked the proper military security clearances to get access to the map of the mine fields. The Navy also refused to give it to him. When the Coolidge approached the channel, a Navy patrol boat was immediately dispatched and radio silence was broken in an attempt to warn the ship. Too late, two mines exploded against her hull, one in an aft section of the ship and one directly adjacent to her main engine room. Mortally damaged, Coolidge's Captain ordered her toward shore and run aground in shallow water. Abandon ship was ordered and lifeboats, rafts, and nets were positioned over the sides of the doomed vessel. Men started going over her sides in a flurry of activity, abandoning all of their personal possessions and all of their military supplies. Though only a handful met with death, many suffered horrible burns as they hit the oily waters surrounding the ship. Just a staging area, Espiritu Santo was poorly equipped to care for the large numbers of injured men streaming ashore.

After wallowing on its side in shallow waters for less than an hour, the Coolidge gave a mighty jerk and rolled completely over on her port side, sinking on the edge of the channel. The shocked members of the 172nd were glad to be alive but were left on Espiritu Santo lacking any of the equipment they would need to get into the war. Months would slip by before they had finally regrouped and refitted to where they would again be a combat ready unit. A small loss when viewed in the overall scheme of the war, I suppose, but a tragedy that marred the military effectiveness and morale of the 172nd for many months to come. What was tantamount to

NAVY DEPARTMENT COMMUNIQUÉ
No.217, December 12, 1942
South Pacific (all dates are east longitude.)

PRESIDENT COOLIDGE LOST ON WAR MISSION; VERY FEW CASUALTIES

The former liner President Coolidge, owned by American President Lines, San Francisco, Calif., chartered and operated by the War Shipping Administration for the U.S. Army, was lost in recent weeks in the South Pacific.

The vessel, operating as a transport, was fully loaded with troops and equipment when it struck a mine and sank.

Through prompt and efficient rescue efforts casualties were limited to four men.

Henry Nelon, 3714 Irving Street, San Francisco, Calif., who is a survivor, was master of the President Coolidge.

The President Coolidge, of 21,936 gross tons, was completed in 1931 at the Newport News Shipbuilding and Drydock Co., Newport News, Va. It was 615 feet 6 inches in length, had a beam of 81 feet 3 inches, and a draft of 28 feet 2 inches.

negligence, however, was the loss of the entire cargo of quinine when the Coolidge went down—an oversight the Navy was quick to sweep under the carpet.

Arthur King, MD, US Army Surgeon, Medical Supply Officer, South Pacific. I was called back to Buttons (Espiritu Santo) when the Coolidge went down to see what could be done to recover the cargo of quinine, since it was the best defense we had in combating the epidemic of malaria running rampant through our troops on the islands. I was fitted out with one of the old Navy diving suits and trained on its use for a couple of days. As it turned out though, a storm came in the day I was going down into the ship and a couple of live mines broke loose from their mooring cables and were floating around the ship. The water got pretty rough around the wreck, making it really dangerous to be in the water with those loose mines floating around. We had to abandon our attempt to get the quinine back. As it turned out, the way the ship was positioned, we would not have been successful in retrieving it anyway.

Today, the final resting place of the Coolidge represents one of the most popular and accessible wrecks of any ship destroyed during World War II. The wreck of the Coolidge is easily reached either by foot or by boat, lying in still waters only a few paces from the shore. Located just a few miles from Luganville on the island of Espiritu Santo in Vanuatu, she is visited by hundreds every year. For those technically skilled in penetration diving, the famous Lady and Unicorn placard can still be seen in its original place of glory mounted just above the mantel in the First Class Smoking Lounge.

Dave Minton, PhM3/2 USN Malaria Control Unit Cactus. Hey listen, those guys with the Army on the Coolidge were very lucky, and I don't mean because they weren't killed or hurt when the ship went down. The best thing for them was what they completely missed out on. They ended up sitting at Espiritu for a year or so until it was all over on the Canal. Later they were sent into New Georgia, but they had a whole year to get ready or soft—one of the two.

Chapter 8

Edging Closer to Victory

November 1942-January 1943

THROUGHOUT OCTOBER AND November 1942, the U.S. Army, Navy and Marines were always one step ahead of the Japanese when supplying their troops. The campaign for Guadalcanal was quite literally being decided now by which side could bring in food and ammunition faster than the other.

After suffering a series of devastating defeats in October at the hands of the Navy and the Marines, the Japanese prepared what they claimed would be the final battle for possession of Guadalcanal during the first week of November. Designated Z-Day by the Japanese High Command, mid November would bring a series of naval battles into Iron Bottom Sound the likes of which the world had never seen nor will ever see again. The Naval Battle of Guadalcanal sealed Japan's fate in the Solomons. Afterwards, they would never again go on the offensive in the Pacific.

NAVY DEPARTMENT COMMUNIQUÉ
No. 193, November 15, 1942
South Pacific

1. Numerous enemy surface forces are active in the southeastern Solomons in an attempt to reinforce the enemy troops now on Guadalcanal and to disrupt our delivery of supplies to our forces which now occupy shore positions in this area.

2. The current land, sea, and air actions in the Solomon Islands are the result of a determined effort on the part of the Japanese to recapture positions in the Guadalcanal—Tulagi area.

3. The engagements are continuing.

Tom Croft, BM3/c, Gun Loader, Turret 1, USS Monssen DD436. That is when they sent us down there to escort several transports with supplies and reinforcements for Guadalcanal. We were providing screening for the operation.

On the afternoon of November 12th, there were a lot of cruisers and destroyers there in the channel. The Japs were throwing everything they had at us from the air. Mostly bombs and strafing.

There was one Jap plane that was still afloat after it hit the water. There were three Japs standing on the wing so we started to maneuver the ship towards them. There was another destroyer with us that was a little closer and the Japs started firing at it with their pistols. The guys on that destroyer opened fire on them with one of their 20mms, ending their lives.

This battle didn't end until late that afternoon.

Later that night, I know we were in formation. Everyone knew the Japanese ships were coming. Our guns were enclosed, our hatches were dogged down. We could not see. You could have heard a mouse run across the deck. Everyone was tense, just waiting.

All of a sudden, all hell broke loose. I think the gun I was on got off two shots when we got hit and the ship went dead in the water. Word came through over the sound-powered phones to abandon ship.

Everyone in the turret left the gun mount and ran to the starboard side to a life raft. Altogether, there were seven of us and one officer. One shell hit right in front of us on the starboard side. We were all hit by it. I am the only one out of the seven who lived.

Robert Michel, PhM3/c, USN Malaria Control Unit Cactus. I remember that night in early November when they issued a Condition Black and that came down to us by word of mouth. Nothing official, just one guy telling another. Condition Red was an air raid, and a siren was sounded. Condition Black was different—the talk was of an invasion. That night we didn't sleep a wink, but sat in our foxholes with rifle and pistol waiting for whatever would come. Thankfully, it never came. But again, we were sort of lost souls as far as the rest of the occupants of the island were concerned. The brass knew we were there but couldn't figure out who we were or who we were supposed to be attached to.

By that time also, we had somewhat lost our conscious fear of dying. We would, of course, avoid it at all costs, but I really think the constant, nerve-racking fear had finally gone. It wasn't until almost sixty years later, when I saw a TV program about the Canal on the History Channel, that I really felt the fear at what happened that night. At the time we didn't know that we on the Canal had been left on our own. We had no idea how precarious our hold on the island was. We could have been wiped out by the Japanese had they gotten through the Navy. But, at the time, we didn't know that.

NAVY DEPARTMENT COMMUNIQUÉ

No.191, November 13, 1942

South Pacific (all dates are east longitude).

On November 12th:

1. At about dawn U.S. naval forces commenced bombardment of enemy positions to the westward of our positions on Guadalcanal Island. Shore batteries were silenced and large fires were started. Seventy-five Japanese landing boats, some of which had been previously damaged, were found at Tassaforonga. Thirty large landing boats were destroyed by ships' gunfire and several others were damaged.

2. At 2:15 P.M., the bombardment was interrupted by an enemy air attack. Twenty-three Japanese torpedo bombers, escorted by 8 Zero fighters, attacked our surface ships in the vicinity of Guadalcanal. Twenty-eight Grumman Wildcats intercepted and shot down 16 enemy bombers and 5 Zeros. Nine enemy planes were shot down by ships' antiaircraft fire.

3. During the above attack a Japanese plane, disabled and burning crashed into the U.S.S. San Francisco (heavy cruiser), killing 30 of her personnel, and damaged the ship slightly. A 5-inch shell from an enemy shore battery damaged the U.S.S. Buchanan and killed 5 of her crew. The San Francisco and the Buchanan were the only U.S. ships damaged in this action. The next of kin of those killed and wounded will be notified by telegram immediately upon receipt of information.

The Naval battle in November will live in my memory forever. I remember that night battle and the Condition Black. We could hear it and we could see the flashes of gunfire reflected off the clouds.

Marine Sergeant, Section Leader 75mm Pack Howitzers I/ 3/11. I call that night in November the Last Stand because that was what it was going to be. We had just three units of fire left for our guns. That's it, just three, and that was nothing. We were told that the position we were in was our last; we were to defend it to the last man. When we had fired off the last of our ammunition, we were told to destroy our guns. We would blow up the breech blocks and make them unusable for the Japanese. Our officers told us that after we had destroyed the guns, it would be up to us to take off and head into the hills. We could fight on like our boys did on Bataan in the Philippines, but we would be on our own. Thank God, it never came to that, but we were prepared for it.

On the morning of November 13th, the Channel off the coast of Guadalcanal was covered as far as the eye could see with oil, debris from the ships, and sailors flailing in the water to stay afloat. Hundreds of bodies littered the oily waters, both Japanese and American. Every small boat that could be mustered from Tulagi and Guadalcanal plied the waters, frantically looking for survivors and pulling them up over the sides of the boats. A continuous line of boats chugged their way from the channel to the shore dropping off men desperately in need of medical attention, turning around, and heading back out into the channel to continue the search.

Virtually every ship in the American fleet had been hit the night before with the exception of the destroyer Fletcher. Burning wrecks littered the black, choppy waters off the coast.

Miraculously, the USS Monssen was still afloat after taking over forty-seven shell hits, some as large as 14" from Japanese battleships. Burning out of control, she floated helplessly in the channel, abandoned except for seven seriously wounded men left behind the night before when those crewmen who were able went over the side.

Tom Croft, BM3/c, Gun Loader, Turret 1, USS Monssen DD436. Some shipmates of mine swam back to the ship the next morning and found seven of us still alive. They took us off the ship and put us on a Higgins boat and brought us to Guadalcanal. I remember I was put in a bomb shelter next to the airport and then put on a plane. Later I was taken by hospital ship from New Caledonia to Aukland, New Zealand, and hospitalized there.

Late that afternoon, the Monssen quietly slipped beneath the waters of Iron Bottom Sound.

Tom Croft was awarded the Purple Heart for wounds he received aboard Monssen that night, and though in good health today, he still carries pieces of his old ship in his back. Years later, Tom Croft would again see his ship when National Geographic released their videotape, "The Lost Ships of Guadalcanal," her guns still played out to starboard just as they were the morning he was dragged off her deck.

Marine Sergeant, Section Leader 75mm Pack Howitzers I/3/11. Those boys—those Navy boys out there that night really saved us. If it weren't for the Navy boys, the Japanese would have kicked us right off that island that night. We looked out the next morning and saw the water covered with all those bodies floating in to shore and realized they were our boys. They saved Guadalcanal.

The Japs Keep Edging Closer…
New Georgia and the Airfield at Munda Point

Located just 180 miles northwest of Guadalcanal lay the island of New Georgia. Sixth largest in the Solomon chain, New Georgia was situated much closer to Guadalcanal than the huge Japanese citadel at Rabaul and, therefore, represented an ideal location on which to construct an airfield. Ideal, that is, if the soil under the overgrown terrain could handle the weight of the traffic. By placing a fighter strip on the island of New Georgia, the Japanese were hoping to tip the balance of air power in the Solomons, giving their fighters the ability to protect their helpless bombers flying missions over Guadalcanal and allow their fighters more operating time over the islands.

While Tom Croft and his shipmates were fighting for their lives in one of the largest naval battles ever to take place, three companies of the Japanese 6th Sasebo Special Naval Landing Force slipped into the surf on the west coast of New Georgia at Munda Point. Landing in the darkness of a moonless night, their orders were to overwhelm the mission located there and prepare to receive engineer and construction units to determine the fitness of the area for a fighter strip. Two engineering units and a rifle company quickly followed, and upon closer inspection realized the Munda Point peninsula presented ideal conditions for their construction plans. Now all the Japanese had to do was hack out a 3,000 foot airstrip right under the nose of the Cactus Air Force operating out of Henderson Field who, by this time, had gained a nominal degree of superiority in the skies over the Solomons.

Thanksgiving to New Years—The Marines are Relieved

Robert Michel, PhM3/c USN Malaria Control Unit Cactus. I have the feeling we started to goof off a little as we got closer to the end of the year. The food did improve. There was more of it and even some variety. Before that, many of our meals were C rations, which got pretty boring. And of course, there was the Spam at whatever mess we ate at; prepared a hundred different ways, but still Spam and anything else the cooks could scrounge up. Strangely, there was an abundant supply of cigarettes that

NAVY DEPARTMENT COMMUNIQUÉ

No. 194, November 16, 1942

South Pacific (all dates are east longitude)

The following resume of recent events in the Solomon Islands area is based upon preliminary reports from the battle area and is necessarily incomplete.

1. Air reconnaissance during the early days of this month revealed a heavy concentration of Japanese transports, cargo ships, and combat units of the enemy fleet in the New Britain-Northwestern Solomons region. An attempt by the enemy to recapture our positions in the Guadalcanal-Tulagi area of the southeastern Solomons was indicated and on November 10th it became evident that the expedition was being launched in force.

2. Japanese naval forces approached the southeastern Solomons from the north as other detachments, including many transports, moved southeastward toward Guadalcanal from Rabaul and Buin, where expeditionary forces had been assembling.

3. General MacArthur's aircraft were of great assistance to our naval forces, both before and during the naval actions. Army bombers made repeated successful attacks on units of the Japanese invasion fleet at Rabaul and at Buin. (These attacks were announced by United Nations Headquarters in Australia.)

4. The spearhead of the Japanese attack was a force composed of battleships of the Kongo class and a number of other vessels believed to have been 2 heavy cruisers, 4 light cruisers, and about 10 destroyers. This unit reached the Guadalcanal area shortly after midnight on the morning of November 11th, intending to bombard

our shore positions prior to a large scale landing from a group of transports which had been observed in the Buin-Shortland area. This Japanese bombardment force was formed in three groups. As they approached the bombardment area they were engaged by units of our fleet and the ensuing battle was fought at close range. It was during this engagement that Rear Admiral Daniel Callaghan, U.S.N., was killed in action. During this furious night engagement the Japanese seemed confused and during the latter part of the battle 2 of the 3 Japanese groups were firing at each other. Shortly thereafter the enemy fire ceased and the Japanese withdrew from the battle and retired to the northward.

5. During the day of November 13th, U.S. aircraft made continuous attacks on damaged Japanese ships which remained in the area. During the late afternoon a large formation of at least 12 enemy transports, under heavy naval escort, headed toward Guadalcanal from the Bougainville area. As a preliminary to the proposed landing an enemy surface force bombarded our positions at Guadalcanal shortly after midnight on the morning of November 14th. Later in the morning, as the transport group drew near to Guadalcanal, it was struck heavily by our air forces and at least 8 of the transports were sunk. The remaining transports continued toward Guadalcanal.

6. During the night of November 14-15, U.S. naval surface forces again engaged Japanese surface units in the Guadalcanal area. Details of this engagement have not yet been received.

7. On the morning of November 15th, four enemy cargo transports were found beached at Tassaforonga, about

> 7-1/2 miles west of our positions on Guadalcanal. These transports were attacked by air, artillery and naval gun fire and were destroyed.
>
> 8. On the morning of November 15th, our patrol aircraft reported the Japanese forces withdrawing to the northward, and no reports of a further action have been received.
>
> 9. During the engagements described above the Japanese suffered the following losses and damage:
>
> a. One battleship sunk.
> b. Three heavy cruisers sunk.
> c. Two light cruisers sunk.
> d. Five destroyers sunk.
> e. Eight transports sunk.
> f. One battleship damaged.
> g. Six destroyers damaged.
> h. Four cargo transports destroyed.
>
> 10. Two light cruisers and six destroyers are the only U.S. naval vessels reported sunk in the actions which were fought on November 13, 14, and 15. The next of kin of casualties will be notified by telegram immediately upon receipt of information.

were always given to us at no cost. Sometimes they were British, sometimes American.

Back in the '30s, when I was in school, there was always a fund-raising drive for the Red Cross to which we, as kids, donated a dime. In those days, a dime got you into the local movie house on Saturdays for a double feature and cartoons, so a dime was a lot of money.

Jump to November 1942, and Thanksgiving on the Canal. We were all standing in line to get our Thanksgiving dinner. Believe it

NAVY DEPARTMENT COMMUNIQUÉ

No. 195, November 19, 1942

South Pacific (all dates are east longitude).

1. Reports just received from the South Pacific reveal that U.S. battleships took part in the action between U.S. surface forces and Japanese surface forces in the Guadalcanal area during the night of November 14-15.

2. Rear Admiral W. A. Lee, Jr., U.S.N., who commanded a task force, which included battleships, has reported that his forced engaged a Japanese surface force in the Guadalcanal area during the night of November 14-15 and inflicted the following damage on the enemy:

a. One battleship (or heavy cruiser) sunk.
b. Three large cruisers sunk.
c. One battleship damaged.
d. One cruiser damaged.
e. One destroyer damaged.

3. It is possible that this report of damage may include some of the damage already reported in Navy Department Communiqué No. 194.

4. Clarification of this point must await the receipt of a complete summary of the action from the commander of naval forces in that area.

or not, real, honest to goodness, turkey with mashed potatoes, stuffing, gravy, and all that goes with it. These Red Cross people, all dressed up in spic and span khakis walked up and down the line of beat up guys handing out one stick of gum per man. The wrapper proclaimed, "Compliments of the American Red Cross"! It took me fifty years to stop hating the Red Cross.

I remember another thing that has to do with the Red Cross. From the time I enlisted, all but about $5.00 a month was going home in an allotment for my mother and brother. While on the Canal, the allotment checks got held up for some reason and my mother went a couple of months without receiving any money. She was worried about me, and not knowing what to do about it, contacted the Red Cross. The Red Cross, in turn, got in touch with one of the officers of our unit who called me in on the carpet. He wanted an explanation as to why I had stopped sending money to my mother. The allotments, of course, were automatically taken out of my pay before I ever saw any money. I told the officer someone had to have made a mistake. He obviously didn't believe me because my advancement in rank to Pharmacist Mate 2nd Class was subsequently held up for months. Another reason I hated the Red Cross for fifty years.

We would occasionally get old newspapers and magazines on the island from the ships' crews. That was the way we got news of the war in Europe, the old newspapers that were sent ashore for us. I don't recall that many of us ever really had any interest in that war. One guy would read what he wanted and then he would pass it on to someone else. Eventually, they would make the rounds and most of the guys would have read them. That's where I saw the ad for the butter I mentioned earlier.

As far as any radio was concerned, I don't believe there was one on the island that was not a military radio. If there was, I sure didn't know about it. I never heard any radio broadcasts while I was on Guadalcanal. I did, however, get hold of an old wind up record player later when I was over on the Floridas. I had one 78 RPM record and it was Bing Crosby's "White Christmas." I played that record over and over again.

I guess it was near the end on the Canal when Dave and I ended up visiting a PT Boat base. Being bored, young, and stupid, we asked if we could go out on one of the boats that was getting ready to leave. We were told OK, but a little later when we were getting ready to board the boat, someone changed their mind and said there had been a change of plans and we wouldn't be able to go. We learned later, whether true or not, that the boat we were supposed to be on was lost that day.

Before I went into the Navy, I was a real regular churchgoer. Once I got in the Navy, I can't recall ever attending a church service. Even on the Canal where I did a lot of praying, I just don't recall any church services. Strange, isn't it?

On November 29, Admiral King, expecting imminent approval by the Joint Chiefs of Staff, sent a directive to Nimitz, Halsey, and MacArthur that Vandergrift's 1st Marine Division would be relieved as soon as possible by the Army's 25th Division, already underway for Australia with their first stop at Guadalcanal.

Japanese grounded troop transport: Guadalcanal

Destined for a long overdue rest in Australia, the 1st Marine Division had put in four grueling months of nonstop combat on a starvation diet, its ranks decimated by the insidious *Anopheles* mosquito and tropical disease.

On December 8, a small flotilla of transports arrived off Kukum, delivering the last of the Americal Division to Guadalcanal. With the unloading operations barely complete, General Vandegrift formally transferred his command of ground forces over to the U.S. Army. Initial elements of the 5th Marines immediately started the laborious process of boarding the now empty transports—destination Australia, and a well deserved rest.

The troops, however, were required to leave the same way they arrived, climbing the ships' cargo nets, falling onto the decks of the ships. Not only were their ranks thinned by months of combat with the Japanese, those who survived carried with them malarial parasites and an array of physical disorders too vast to catalog.

For many, the ordeal of reaching the sanctuary of a ship's deck proved too much. Many had to be lifted and carried aboard by healthier comrades and sailors. Many years would pass for the survivors before the final effects of the *Anopheles* and Guadalcanal would become a memory. Many more would carry the scars of the island's pestilence for the rest of their lives.

Marine Sergeant, Section Leader 75mm Pack Howitzers, I/3/11. When I got to Guadalcanal on August 7th, I weighed 168 pounds and was in great shape, real fighting trim. When I left on December 18, barely four months later, I weighed 127 pounds. That's how much weight I lost.

We were taken out to the transport ships and they threw the cargo nets down over the side for us to climb up. None of us could do it. The sailors had to lower the gangway instead. The gangway was the stairway or ladder that the big shots used to climb aboard a ship. We had to climb up the gangway and the sailors then helped us off and onto the deck of the ship. We couldn't make it up the side on the cargo nets the way we did when we came in four months earlier.

PFC Stewart Winterton, Marine Rifleman B/1/7. I was carried down off the ship on a stretcher and went straight to the hospital, where I was in and out of until May 1943. Then I was shipped back to the States. I had malaria really bad. I would almost get over it and it would come back.

Robert Michel, PhM3/c, USN Malaria Control Unit Cactus. I do remember when the Marines pulled out. I don't remember that it was such a big deal for us, though, to leave. By that time, things had slowed down quite a bit and what Japs were still on the island seemed a long way from where we were. I don't think at the time we were even aware that the Army had taken command of the island. It just didn't really affect us.

John Francis Richter, PhM3/c Corpsman USN, 75mm Gun Half—Track, Headquarters & Service Battery, 11th Marines. Yes! I got malaria, but not on the Canal. I hadn't contracted my first of four malaria attacks until arriving at Brisbane, Australia. I strongly believe the Atabrine tablets I took on the Canal suppressed the symptoms until I had stopped taking them in Brisbane. One minute you were burning up with high fever, the next moment you were thinking your front teeth would chip as you reversed into chilling. Fever, chills, fever, chills; that was the going routine.

William C. Dickman, MSgt. USMC Enlisted Navigator/Naval Aviation Observer VMJ-253 MAG-25, South Pacific Combat Air Transport, SCAT. I contracted malaria in April 1943. At the time I thought I had caught it on the Canal. I was flown to a hospital for recuperation at Rose Bay, Australia. I was there for three weeks. The Atabrine didn't really seem to help me much, though it did turn me as yellow as a Chinaman. I continued to get malaria attacks every year for ten years, up until 1953; since then, none. The mosquitoes that were responsible for spreading the malaria were really playing hell on all our troops up there.

On December 9, the day after our Marine Sergeant crawled aboard the troopship that was to take him to Australia and a much needed rest, persistent photoreconnaissance of New Georgia's Munda Point convinced U.S. military leaders that the Japanese were indeed scrambling to complete construction of an air base on its western coast. Alerted weeks earlier by Major Donald Kennedy, a commissioner of the old British Protectorate who was now a member of the Island Coastwatchers Service operated by the Australian Navy's intelligence division, the Japanese had cleverly camouflaged their work. With their engineering crews working under the cover of darkness, the Japanese had suspended the tops of the coconut trees on heavy wire, cutting the trunks, and leaving the treetops dangling in midair from the cable. Viewed from the air, Munda Point appeared unchanged and as pristine as it had for hundreds of years. This gave the Japanese, undisturbed by the Cactus Air Force, more than three weeks to complete two crushed coral runways underneath the false canopy of treetops. Heavy defenses had been installed. Antiaircraft positions circled the strip and additional Japanese infantry units had completed strong defensive positions. In just six short months, Dave Minton would see firsthand just how deep the Japanese had dug themselves in.

As the ground campaign on the island began winding down and becoming more manageable in December 1942, Marine combat units were gradually relieved and shipped off to Australia. Malaria Control Unit Cactus, however, stayed on at the rear area bases on Guadalcanal, Tulagi, and the Florida Islands expanded to accommodate the men, equipment, and supplies needed further up the Solomon chain.

Though it may seem that the campaign for the Solomons was over, more Japanese strongholds still lay within the archipelago. New airfields at Munda Point on New Georgia and Vila River on Kolombangara posed a huge threat to both the security of the bases left on Guadalcanal and future Allied plans to encircle and annihilate Rabaul.

Just as Marines were pulling out of the lines on Guadalcanal and the Army was moving in, preparing for its final assault on remaining enemy strongholds, the first wave of Japanese fighters dropped into the airfield at Munda Point. The Cactus Air Force was ordered to respond on December 23, launching the first of continuous, daily strikes against enemy bases located in their backyard. Within a matter of weeks, US air raids had rendered Munda untenable to Japanese fliers. Though maintained in operable condition by Japanese ground crews over the coming months, the airfield provided little more than emergency runways for damaged or crippled enemy planes. Once again, air power alone proved damaging yet inconclusive—Japanese forces still held the strategic western shores of New Georgia and it was determined the only way to get them out was to forcibly remove them with Army and Marine ground units.

Thomas Gross, US Army 97th Field Artillery Battalion, Americal Division. I was still in New Caledonia at the end of 1942—I had a lousy New Year's Eve and Christmas wasn't worth a damn either. This was the first time in twenty four years I hadn't been home or seen snow. It was hotter than hell, and when it rained it seemed the gates just opened and out it poured.

Right after New Year's, rumors started flying that we were soon to be leaving New Caledonia. Some thought we would be headed to New Guinea, India, or the Solomons. Some souls even believed, or perhaps hoped, it would be Australia or even New Zealand.

On January 11, we finally boarded the USS McCawley minus our mules. At the time, we thought sure we were headed for the Solomons. There were Air Corps, antiaircraft, and even a Naval Construction Battalion on the same ship with us. After laying outside Noumea harbor all next day, the 12th, we finally said goodbye to the Rock. We had been on that hole for almost ten months, but I had a sneaking suspicion there were worse places in the world.

We arrived at Guadalcanal on January 16, 1943, about 0630 in the morning. We landed at a beach and immediately started unloading our equipment from Y-Boats and Tank Lighters.

I always thought New Caledonia was hot, but here you would sweat until it just dripped off you just by doing a little work.

William C. Dickman, MSgt. USMC Enlisted Navigator/Naval Aviation Observer VMJ-253 MAG-25, South Pacific Combat Air Transport, SCAT. Espiritu Santo was one of our main hospitals in the area and the island of Efate in the New Hebrides was a large Army Medical area. This is why, when we were scheduled to fly to Espiritu, code named Buttons, or Efate, code named Roses, from Guadalcanal, we were hauling wounded and mail and also bringing out flight crews stationed there for rotation.

The bases we operated out of changed code names many times. The code names of the islands were stamped on cruise boxes, gas drums, bombs, torpedoes, and other equipment. This helped to divert the equipment to the right locations.

We normally picked up mail and personnel when we left New Caledonia, Epic, and stopped at Espiritu to take on gas in 55 gallon drums, bombs, and small arms ammo. Many times our flight crews would go direct from New Caledonia to Guadalcanal, code named Cactus, if wounded had to be evacuated immediately or the need for supplies was great.

This was a tricky flight as we had to plan on arriving just at daybreak before Japanese air activity picked up. Also, we could not land at night because the flare pots set out along the runways would play the devil with our blackout procedures.

All of the fighter, dive bomber, and torpedo aircraft that we led for navigational aid were either from the fighter strip or the bomber strip at Espiritu. We always had to conduct this type of operation in the daylight in the event that someone might have to ditch on the way. All of our aircraft were unarmed in order to keep them light and to carry more fuel. This was a four and a half or five hour flight over shark-infested waters, and the last two hours of the flight we were exposed to possible enemy aircraft attack.

Our navigation had to be very precise in the event an aircraft ditched. A PBY squadron would then be contacted by radio in the event of a splashdown, and we would radio by Morse code the latitude and longitude of the downed pilot. We did not lose a single aircraft in our eight month tour in the area. We escorted Marine, Army, Air Force, and Navy aircraft using these navigational procedures. Due to the range and the weather in the area, we did not lead tactical aircraft direct from New Caledonia to Guadalcanal since this flight was six and a half to seven hours in length.

We also conducted flights to Auckland, New Zealand, and Sydney and Townsville, Australia, to carry flight crews and wounded either on rotation or R & R.

In the first few months, the squadrons involved in these operations were VMJ-253, VMJ-152, and one Army Transport Squadron. Later, of course, there were many other squadrons involved. Squadron VMJ-253 set a record on September 1, 1942, for the longest nonstop mass unit flight on our way to the Solomon Islands campaign.

Robert Michel, PhM3/c USN Malaria Control Unit Cactus. I remember one of the flights in particular. I was either coming back to the Canal or leaving it in an empty C47 that was being used as an ambulance plane. The plane had stretchers fastened to the side of the fuselage behind the cockpit. These were placed in the plane end to end, two high, running down both sides of the plane. Since all the stretchers were empty, I can only assume we were leaving the Canal and heading to another island to pick up casualties, where we would be filled again for the trip back to Henderson Field.

I was in the cockpit talking with the pilot and having a cigarette. The copilot was asleep in his seat and had been since I got aboard.

We were about five thousand feet up and on a steady heading when the pilot asked me if I flew. Could I fly a plane? I said yes, but of course, I couldn't. He promptly informed me he was going back in the plane to take a nap on one of the stretchers and that I should stay on the compass heading at this altitude. With that the pilot got up and I sat down—I actually flew the damn thing. He told me to wake the copilot when I saw land as that would be the Canal. You talk about white knuckles, brother, I had them. What in the hell have I gotten myself into this time, I thought. I never flew a plane in my life. But after awhile I started fooling around a little bit, up a thousand, down a thousand, a little to the left, a little to the right. I eventually saw the island and woke the copilot. He was a little groggy and couldn't for the life of him figure out who in the hell I was. Remember, he was asleep when I got on the plane. To this day he must still be trying to figure out who in the hell the guy in the left seat was.

Dave Minton PhM3/c, USN Malaria Control Unit Cactus. I think Bob is full of it about flying the plane. I don't really know, though, I wasn't with him at the time. I had already been transferred out and was with another unit. It sure sounds like the stuff Bob used to do, though. He had no fear. He was always doing something crazy, and flying the plane is just the type of thing he would do.

Thomas Gross, US Army 97th Field Artillery Battalion, Americal Division. We were set up only about six miles from the "front lines" and could see where the shells from our 105 and 155mm howitzers fell on Jap-held land. The area we were in held the only worthwhile thing on the whole island—Henderson Field, or as we called it, "the bomber strip." There was also a newly constructed fighter strip made of concrete about a mile or two from the bomber strip. During the day, the air was filled with P38s, P39s, and once in a while a few P40s. There were lots of Grumman Wildcats and Avengers. These planes made a terrible racket. The P38 Lightning, or split tail, was the only plane that didn't rattle your brain as it flew overhead at low altitude. The Lightning was a really sweet plane. There were also quite a few B17s and B25s there, and once in a while we would even see a DC3, the plane all of the commercial airlines used back home.

We weren't there long until we started hearing a lot of the tall tales from the Marines about their fighting the Japs. But, the Marines were really a fighting bunch of men—they took no prisoners whatsoever. If they captured a Jap, he was shot on the spot. The Army, on the other hand, wanted to take prisoners and end the fighting, but the Marines just wanted to keep going on. They really put the fear of God in them. It got to the point where a Jap would surrender to the Army but never to a Marine.

The Jap position about this time also looked pretty hopeless because they couldn't get any supplies. It was rumored that some dead Japs were cut open and were found to have been eating grasshoppers, roots, and the like. They no longer rushed one of our soldiers with just a sword, nor did they chatter and smoke like they used to in order to draw our rifle fire to learn our positions. A well-placed hand grenade didn't make any flash when it landed. There was even talk that we could take the island any time we wanted to now, but lots of our men would have been killed, so we just held our lines waiting to starve the Japs into surrender.

The Japs did raid us every night with their airplanes, though. One was known as "Washboard or Maytag Charlie." When he came over, he always stayed up at about 3,000 feet or so. He kept us up night after night. It was damned miserable to stay up all night and then have to work all day afterwards. Whenever we had night fighters up he would stay away, but that wasn't very often.

I was already sick of the place, but always figured that in six to ten months we might be relieved, since I didn't think we could stay in the tropics like this forever. It seemed it would be worth any price if we could have been sent to some civilized country or even home. The island was very damp and it rained about every two or three days. There were beautiful palm trees and good cocoanuts everywhere, but a full moon didn't look all that romantic through the waving palm trees when I was sitting there waiting for "Washboard Charlie" to dump his load. I swore at the time those boys who wrote the travel brochures had never been west of Jersey for my money. Beautiful South Seas islands—Nuts!

I often wondered how I would feel when death put out her hand to me. Well, I found out one night at the end of January when we had an air raid. It was about 9:30 P.M. and we had our first air raid alert of the night, followed by the "All Clear" about an hour later, around 10:30 or so. At about 11:30, we got another warning, and then the planes started coming in. My buddy Tuma and I were in our foxhole and were watching the searchlights and AA guns open up on three planes. As long as I was able to look at them from an angle I was OK, but when I had to look straight up at them, I ducked. Right about this time, Charlie swept his plane of everything, including "Tojo's slippers." The first bomb hit about 500 yards to the west of our positions and the rest started coming right down the line towards us. My first reaction was that I could clearly see a bomb drop right in a dugout and throw about six of our boys in all directions. As the bombs kept coming closer, I was scared that one might drop close to me and bury me with dirt. As the closest bomb dropped, I tried very hard to hear the shrapnel whistle through the air, but couldn't. A Jap bomb doesn't whistle or scream but rather "swooshes." One of the bombs dropped about six feet from the dugout that the hitchin' crew was in and it tore the officers' tents to shreds. It started a fire in the white gas drums which set fire to the cooks' tent. No one was killed and only one man was hit with a small piece of shrapnel. Some of the men's reactions were really amusing—the bombs made a lot of Christians that night.

At the end of January it was rumored that the "Pets," or mules, were due to arrive within a day or so. Later that day, their arrival date was changed to a couple of days later. We knew that when the mules got there our troubles would really start. I'll never forget how we had been told we would never leave the States with those mules—Phooey!

When the mules finally did arrive a couple of days later, it took two whole days to get them off the boat, and a third day just to unload the hay. One could always live a peaceful life when you didn't have to mess with those mules. My swearing would always jump about 99% when I was around them.

As soon as we got the mules in, we left for the front. I thought that since we were in the artillery, we would stay in back of the lines—Hell! We ended up behind the Japanese lines. We moved our guns up in some hills that were parallel to the Jap lines. We did have infantry protection on our flanks and out in front of us. Still, snipers gave us some trouble the day we moved into position—it was a helluva mess for awhile. Some of the younger boys just threw away their halter shanks and the poor mules were wandering around out there all by themselves.

Every once in a while we passed a place that smelled just awful and knew there was a dead Jap. But it was a lot better that they were dead than alive. The ants and flies were just about driving us all crazy and snipers were always taking pot shots at us. We walked all day with wet feet and mud over the tops of our shoes—nothing but rain and misery. It seemed like it was only about 120 degrees there.

By January 1943, Navy and Marine brass were convinced the men of MCU Cactus had learned their job well on Guadalcanal, having gained valuable experience seeking out breeding grounds and identifying deadly *Anopheles* larvae. In an effort to make South Pacific Island bases safer and more hospitable for the growing number of troops flooding the area, Malaria Control Unit Cactus was ordered to expand its area of operations beyond the perimeter and rear areas of Guadalcanal. New orders were issued stating that malaria control units would now be an integral part of all future Marine and Army landings as offensive operations advanced U.S. forces up the island ladder toward New Guinea. Malaria Control Unit Cactus was to stay in the war in an expanded capacity.

MCU Cactus Guadalcanal, January 1943
Bob Michel standing, back row far left, with pith helmet
Ensign Kenneth Knight kneeling, front row, second from right
Chief pharmacist mate Rountree kneeling, fourth from left

Marines were being relieved on Guadalcanal and readied for their next offensive operation against the Japanese. Since Guadalcanal was now an Army show, the original members of MCU Cactus who had been on the Canal for five months were being reassigned to other areas under Navy control, a handful at a time. The first week in January, Dave Minton became one of the first to go, assigned to the island of Espiritu Santo in the New Hebrides. It was only about a month later that Robert Michel was packed off to the Florida Islands, first Tulagi and a short time later to Nggela Pele. Having served together since the fall of 1941 in the States and surviving five months of constant nonstop enemy shelling and bombing, it would be almost another ten months before Dave and Bob would cross paths again.

February, My Job Here Is Done

Thomas Gross, US Army 97th Field Artillery Battalion, Americal Division. On February 7, our unit moved up nearer to the front lines. They kept telling us the Japs were on their last leg and that some U.S. units had landed farther up the island and were driving back towards us. Naturally, what we heard was all rumors, but the best rumor going around at the time was that everyone on Guadalcanal would be back in the States by Easter. I thought we'd be lucky to see the States by 1945!

Anyway, it wasn't but a couple of more days when on the 10th, a phone call came in to our unit. We were told the Japs had finally given up the ghost and that the island was finally secured. Again the rumors flew and we heard that there were upwards of 100 Jap prisoners taken. All of the boys started wondering where we would be sent from here. I thought we would be staying for quite some time.

One night just before we heard the island was secured, my two buddies and I were lying in bed almost going to sleep when all of a sudden I heard a trickle of water. I looked up to see water running through our tent. Before long it was a foot deep and all hell broke loose. My buddy, King, later picked up a dead Jap by where we had our pillows. He picked him up and tossed him in the bushes. The entire next day we spent drying out our equipment. Lots of our boys lost equipment that night in the flood which they never saw again.

Robert Michel, Phm3/c USN Malaria Control Unit Cactus. All in all, I was on the Canal until it was finally secured in February, 1943. Then I was sent to Tulagi for a couple of months in early March and then to a Marine radar outpost at Siota on Little Florida

Island (Nggela Pile) in May. Dave and I were never together again in a unit after the Canal.

While on the Canal, we used local natives for laborers. I made friends with a couple of natives, so I guess that's the reason I got the job recruiting over on the Floridas. The brass figured I knew how to get along with them and would, therefore, be able to recruit them for helping with Malaria control. I was also able to communicate with the natives in their Pidgin English.

When Japanese troops started moving into the Solomon Islands in May 1942, first on their order of business was to let the natives know the British were no longer in charge. There would be a whole new set of rules and regulations the natives would need to know and obey implicitly. Ever cautious of outsiders, the natives initially took a wait and see attitude toward their new masters. Perhaps being members of the Greater Asia Co-Prosperity Sphere would be better than the stiff colonial rule of the British. At least the new rulers were Asian.

It didn't take long, however, for Solomon Islanders to start feeling the weight of the conqueror's boot. Within weeks, the Japanese showed their true colors by raiding the islanders' gardens, treating their crops as if they were the property of the Emperor. Next, the Japanese issued a proclamation that all male islanders between the ages of fourteen and fifty years must immediately report for work. The Japanese decree stated that all laborers who performed their work to the Emperor's satisfaction would be provided food and, after one month's service, would be issued a certificate stating they were legitimate civilians in residence on the island—whatever that meant.

Right about then, native leaders realized that if they were to ever win possession of their islands back from the Japanese again, they would have to set aside old rivalries and stop the intertribal warfare that had plagued their lands for so many hundreds of years. Ironic that it would take an invasion by the Japanese for them to finally realize the unity and peace promoted by the British for over forty years. And maybe, when the war was over and the Japanese were finally gone from their homeland, the Americans just might side with the Solomon Islanders and kick the British out as well. At least that's what many of the island leaders hoped.

Regardless, never was there a more loyal and trustworthy indigenous population than were the dozens of tribes living in the Solomons. Organized and led by such stalwarts as Jacob Vouza, decorated by the Marines for his bravery and later knighted by Queen Elizabeth for years of loyal service, islanders quickly volunteered as scouts, dockhands, porters, and guides. Anything the Americans asked of them, islanders enthusiastically obliged. If malaria control and the hot dirty work that it entailed was part of the package, so be it. Natives not only worked side by side with Bob and the men that followed, spraying diesel oil on stagnant water and digging drainage ditches, but also made the Americans' time on the islands as comfortable and safe as possible.

Chapter 9

Like Robinson Crusoe

February 1943-May 1943

AT THE END OF February 1943, when Bob and the remaining members of MCU Cactus were transferred from Guadalcanal (which was now under the command and control of the U.S. Army) just 20 miles across Sealark Channel to Tulagi, they were once again in territory controlled by the Navy. In early 1943, Tulagi, Gavutu, and the southern shores of the Floridas were a beehive of activity encompassing the busy harbor of Purvis Bay, a PT Boat base, a small boat maintenance and repair facility, and huge shore based supply depots. Though the campaign on Guadalcanal had captured the attention of the world for the past six months, it mustn't be forgotten that Tulagi wasn't always so peaceful.

When the war broke out in 1941, the tiny island of Tulagi, located just twenty miles north of Guadalcanal (and south of and adjacent to the Florida Islands), was the capital of the British Solomon Islands Protectorate under the management of the Australian government. Tulagi, with its large, natural deepwater harbor, would actually represent a much greater military significance to the Japanese than did the more famous island of Guadalcanal.

The Japanese Yokahama Air Group, therefore, occupied the island on May 3, 1942, sending a total of 880 naval personnel and Korean conscripts to begin work on a seaplane base. Their plans were to later expand the harbor and port facilities to handle Japanese Naval fleet operations in the lower Solomons in their push toward Australia and New Zealand.

Before we pick up with Bob's story in February 1943, let's take a moment to step back in time just six short months to that historic day in August when the United States launched its first offensive of the war in their bid to halt Japanese expansion across the South Pacific.

Tom Croft, BM3/c Gun Loader, Turret 1, USS Monssen DD436. Well, we all knew something was going to happen that morning. We really didn't know where we were going. Heading into the islands there, on August 7, 1942, they gave us chow early and then sent us to our stations; General Quarters. If I remember it right, we started shelling Florida and Tulagi shortly before 0630. We kept the shelling up for a while even when the Marines started their landings.

There was a seaplane sitting out there in Tulagi harbor that morning. There were some Japanese that were trying to swim out to it. We fired one salvo at that plane and it was blown to bits.

One of the captains of the 1st Marine Division that went in early that morning on the Higgins Boats was killed. When they picked up his helmet, it had a bullet hole square in the middle of the forehead.

Right on top of a hill on the island there was a machine gun nest. They started firing at us on the ship. We blew it up. Not long after that the Marines had the island secured.

We stayed at General Quarters all day that day. By 1700 that evening, we had the islands secured.

Lt. Cmdr. Lewis P. May, USNR Ret., Assistant Supply Officer, USS President Hayes APA-20. It was calm where we anchored at Tulagi that morning. Our unloading was halted, though, when we heard of fierce fighting ashore. During the night, the situation was obscure, but serious. Then we got the news that enemy ships were on the way, and although we still had the major portion of supplies and provisions for the Marines still on board, we were ordered to pull out unescorted. So we left at top speed. I think it was almost a month before we were able to return with supplies and food by which time the Marines were eating Japanese rice and K rations.

William Ricciardi, MM2/c Boat Gang USS Crescent City. A real good buddy of mine was killed over there on Gavutu, not too far from Tulagi. He was a Marine and Jewish. His name was Brotsky but he changed it to Sterling. Barnie Sterling was his name, and he is buried there with about five or six other Marines who were killed there that day. He was buried right behind one of the buildings.

I had another good friend who went in that morning on one of the boats. They were coming into the beach and there was a short wall in front of them just off the shore that was maybe two and a half to three feet high. As they dropped the bow and the Marines started running out of the boat, a bunch of Japs rose up from behind that wall with a machine gun and shot all of them, including the coxswain of the boat. The only one who survived was a young kid who was the bow hook. He was so scared he crawled back to the boat controls and just rammed that thing in reverse. The boat was dragging the bow across the water because he never even raised it. He just kept backing out of there. He was about five miles out when a destroyer on picket duty picked him up. If it hadn't have been for that destroyer stopping that kid, he probably would have backed all the way to the States, he was so scared seeing all those guys killed.

At 0800 on the morning of August 7, 1942, over eighty U.S. ships gathered in Sealark Channel to launch the first Allied offensive of the war in the Pacific against Guadalcanal. Marines of the 1st Raider Battalion, under the leadership of Colonel Merritt Edson, stormed ashore on Tulagi raising the American flag over the island later that afternoon.

After being secured by Marine Raiders, tiny Tulagi and nearby Nggela Sule and Nggela Pile became part of the Pacific supply chain for the rest of the war, providing an excellent base for PT boats, seaplanes, and a safe harbor for ships in the area. By February 1943, when the island of Guadalcanal was finally secured and members of MCU Cactus were transferred to other locations, Tulagi and Nggela of the Florida group were home to almost 6,500 U.S. troops stationed there in service and support capacities.

Robert Michel, PhM3/c USN Malaria Control Unit Cactus. I got over to Tulagi by boat. There were only about four or five of us left from the original unit at the time. Our quarters were much better there than on the Canal. With wooden floors for our tents it was paradise. The food also was much better.

William C. Dickman, MSgt. USMC Enlisted Navigator/Naval Aviation Observer VMJ-253 MAG-25, South Pacific Combat Air Transport, SCAT. As far as I know, we had no control of the malaria problem in Tontouta, New Caledonia, at the beginning. In January 1943, though, the following precautionary steps were ordered. All tents that had their sides rolled up had to be rolled down daily because water running off the tents would collect in the rolls and allow mosquitoes to breed. All standing water was drained. When flight crews attended a movie, the control group passed out mosquito repellant for use during the show. We always slept under mosquito netting each night. All flight line crews wore mosquito nets over their helmets during the day. This was the standard routine until the Malaria Control Unit started spraying twice a day, morning and evening. All loose brush was also cleared away.

I wonder, now, if the mosquito control people knew at the time that there were no malarial mosquitos on New Caledonia? They sure did take all the precautions, though.

Though Japanese troops had successfully evacuated almost 11,000 surviving troops from Guadalcanal early in February, and effectively ceased any further ground action on the island, they were by no means ready to call it quits in the Solomons. Hundreds of Japanese planes, scattered across airfields on the occupied islands of Rabaul in New Britain, Bougainville, and Shortland, were still within flying range of American bases on Guadalcanal and adjacent waters. A series of massive air raids were scheduled to disrupt Allied supply efforts and stall any intended advance on Japanese positions further up the islands. The first strike of Operation I was, therefore, set to launch on April 4, 1943. Subsequent assaults were to follow every two days for the next week. The first strike on April 4, however, had to be scrubbed due to bad weather over the islands, pushing the launch of Operation I back for three days.

On the morning of April 7, Japanese planes finally took to the air heading straight for Guadalcanal and for any ships that looked a likely target in the vicinity. Over the next four days, a total of 486 fighters, 114 carrier based bombers, and eighty land based attack planes took part in the operation. The first attack swept in over the west coast of Guadalcanal and headed straight on to Tulagi and Allied transports clogging Purvis Bay. Follow-up raids continued over the next week with raids hitting the island and nearby ships on April 11, 12 and 14.

Robert Michel PhM3/c, USN Malaria Control Unit Cactus. In March or early April 1943, while I was at Tulagi, there was a large task force forming in Purvis Bay for a big push north to another one of the islands. Anyway, this task force of ships was sitting in the bay and we had an air raid. One of the ships, a tanker, was hit and the bay erupted in flames from the oil and gasoline. Many bodies and survivors were brought ashore and up to a first-aid tent that was right next to ours. It took me many years to forget the sound of the screams of pain that I heard. Men were horribly burned. Skin burned completely off, and corpsmen and doctors trying to do what they could with what they had in the tent. I am sitting here today holding back the sobs.

Thomas Gross, U.S. Army 97th Field Artillery Battalion, Americal Division. On April 7, 1943, about 10:30 in the morning, the "bogies" came over. From all reports, there were about one hundred of them. They dropped most of their eggs on Tulagi, right across the channel from us, and supposedly hit a small oil tanker. We heard that they shot down two of our P38s on the way in, right over Savo Island, even before they got into the fight over Tulagi. We were supposed to have shot down twenty-seven Nips that morning in the fight.

The air attack that took place on April 7th that Bob and Tom have described for us caught and sank the old oiler, USS Kanahwa. Abandoned almost immediately upon being hit, the crew of the Kanahwa suffered heavy casualties. The destroyer Aaron Ward took several hits in the first attack that morning as well, but was able to stay afloat for several hours before finally going down.

Nggela Pile and Siota Mission

Situated just north of the island of Gudalcanal, and within sight of Honiara, the present day capital of the Solomon Islands, lies a small group of volcanic islands called the Florida Island Group. Most visible on any map you may reference of the South Pacific today are the two largest islands of the Florida group called Nggela Sule (pronounced Neela Soolie) and Nggela Pile (pronounced Neela Peelie). It was to Nggela Pile that Bob was transferred in May 1943, following his short stay on Tulagi, and to which he refers in his continuing narrative whenever he mentions the Floridas.

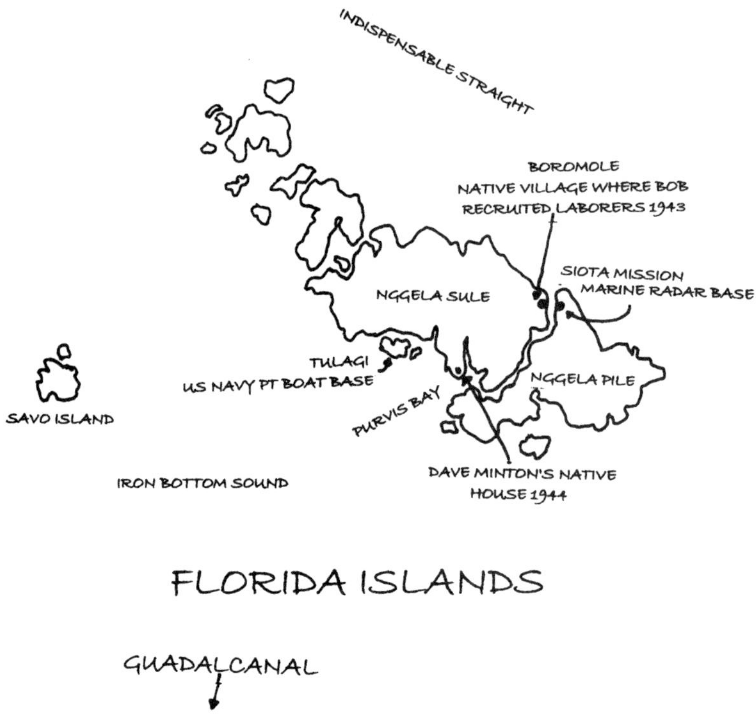

The two islands, Sule and Pile, are separated by a narrow channel which is called the Utaha Passage. It was up this waterway that Bob traveled north from Tulagi by Higgins boat to the northwest shore of Nggela Pile. He landed at the Marine radar station, code named Begonia, located at the mission village of Siota. It was the Utaha Passage that Bob later crossed when visiting the native village of Boromole, the home of his houseboy Graves.

Robert Michel PhM3/c, USN Malaria Control Unit Cactus. When I left Tulagi, it was by Higgins boat. You can see on a map that there is a passage that separates the Floridas into two islands. We sailed up that waterway to the far end and I was put ashore on the smaller of the two islands, Little Florida, just to the north of Tulagi. A village called Siota is exactly where I was. At the very end of that island is where the Marine radar detachment was stationed, which was about 100 yards north of the village. My duty there was independent duty to the Marines and malaria control.

Florida Islanders recruited in the fight against malaria
Ngella Sule, the Floridas 1943

My job, also, was to recruit native labor to help me with malaria control efforts by ridding the island of mosquitoes. To the south and along the edge of the water is where most of our drainage work was done. As I remember it, the area around the Marine radar station had no standing water at all. Most of the water and swampland, each a primary mosquito breeding site, was to the south of Siota and inland.

Brian Taylor, PhD, Nottingham, UK, Onetime Government Entomologist Malaria Eradication Programme, Solomon Islands. Siota Mission, if I remember rightly, I stayed there on my first Solomon's field trip. It was a United Church of the Solomons, essentially a Methodist mission.

I traveled all over the place and remember discussing with a then old man how the Americans had used aerial spraying all over north Guadalcanal and the adjoining islands to kill all the mosquitos and, as it turned out, all the birds.

Robert Michel, PhM3/c USN, Malaria Control Unit Cactus. I don't remember too much about the Marines at the radar station there except that they had a tank, and I used to take rides in it. What always impressed me then, and what I still remember, is how smoothly it rode. My first impression of a tank was that it would be a very rough ride. That proved untrue the very first time I rode in it.

Guadalcanal, Florida, and the rest of the islands in the Solomons, are populated by a happy, smiling race of Melanesians. Extremely dark of skin with light, almost tawny hair, the islanders lived off the land and the abundance of the surrounding ocean. With strong family ties and a society steeped in ancestral tradition, the indigenous natives had been swept from the Stone Age into the modern age by the invading armies of the United States and Japan.

The islanders from the very onset of hostilities, caught up in the destruction and violence of the war, remained staunchly loyal to the Allied cause. Not once, throughout the entire war in the South Pacific, did native islanders ever betray an Allied position to the Japanese. Whenever U.S. or Allied patrols or scouts moved through an area, they moved amongst friends. It was not uncommon, in villages that had experienced Japanese oppression first hand, for Solomon Islanders to rise up against their enemy and defend their families and homeland with whatever weapons they had at hand.

Brian Taylor, PhD, Nottingham, UK, Onetime Government Entomologist Malaria Eradication Programme, Solomon Islands. Curiously, perhaps it was the "colonial ethic," we never really considered "tribes" per se. I was well aware of the great diversity of language, etc. and the inter-island rivalries, but the names of the tribes were not used in our work or employment of the locals.

It is not entirely true that the locals remained loyal to the Allied cause. Horton, in his book **Fire over the Islands**, which describes the activities of the "Coastwatchers"—mostly government officers, like himself, who had stayed on when the Colonial administration was evacuated and gave much early warning aid to the Allied Forces—wrote how there was a significant group sympathetic to the Japanese. That group saw no benefit from the old colonial "Protectorate" and thought the new liberators (the Japanese) would be a better bet. What Horton saw as the turning point was when the Japanese chose to destroy the gardens (slash and burn agriculture) of uncooperative locals on Guadalcanal. From that time on there seems to have been no sympathy of any sort towards the Japanese—rather to the contrary, as the locals were very efficient killers, with superb bush skills.

Robert Michel PhM3/c, USN Malaria Control Unit Cactus. When I first arrived on Florida, I moved into a house that may have been a missionary's house before the war. It had been shelled by us or by the Japs and was pretty much destroyed except for an enclosed porch. The natives rebuilt the porch and screened it in with mosquito netting, and that ended up being my living quarters. I had a cot and a table, on which was a Primus Stove. That was it.

I didn't actually live with the natives over there. The natives lived nearby in a temporary village they had constructed for themselves. A number of them would go "home" at night and then come back the next morning for work. Their home was just a relatively short paddle across the channel between the two islands from where I was living. Their permanent village was somewhere in the vicinity of, but just south of, the present day village of Boromole.

Children of Boromole Village
across the channel from Siota Mission
Ngella Sule the Floridas, 1943

I was able to catch rain water from the roof of the house and I used it for drinking water, bathing and laundry. I had the natives rig two fifty-five gallon drums, that we had scrubbed out, just below the gutters of the roof, one on the sunny side of the building, the other on the shady side. When it rained, the drums would fill up with water. The drum on the sunny side would have warm water, the other cool. I used a gravity fed hose to bring the water from both drums into the house and fastened a tin can with holes in it over the ends of the hoses. By gosh, I had hot and cold running water for a shower. My shower on Florida was like paradise—soft rainwater.

I already had a couple of boys with me, and one named Graves (pronounced Grav-is) Malana became my houseboy. He spoke some English and was able to communicate for me to the rest of the boys.

The adult male natives—those fourteen years of age and up—that we recruited, lived with me on Little Florida near my house. I kept a cadre of about eight or ten with me on the island to do the hard labor of spraying. The rest were sent back to Tulagi or Big Florida. Most of them were all from the same tribe from Big Florida, although I had no idea what the tribe was called. I do remember they didn't seem all that friendly toward other tribes in the area. They weren't warlike, just not that friendly when it came to other natives.

While helping me with malaria control, they performed any manual labor necessary, like making holes in the marshy areas with poles that I would put dynamite in to create a drainage ditch to the ocean. They also used machetes to clear trails wherever we were doing the spraying. Of course, Graves, my houseboy, was exempt from all that and acted as my sub-chief. I was the chief, "Missa Bob." Graves also did the basic recruiting and kept my house for me, cooking my tea and food on his Primus stove.

I had been supplied with a bag of shillings which at the time, I think, were worth nineteen cents each American and a bale of tea. I paid the natives one shilling a day along with a handful of tea.

Graves and I would go across the water separating the two Floridas to the native villages to recruit laborers. I would then have them come back across the channel to our area to live and work.

Brian Taylor, PhD, Nottingham, UK, Onetime Government Entomologist Malaria Eradication Programme, Solomon Islands. My first visit to Siota Mission Station was on my first "tour" in the Solomons and so was something of a confused haze for me. From that haze, though, I remember there were certainly still drainage ditches in the area and the house Bob Michel describes quite probably was still within the same compound as earlier.

Robert Michel, PhM3/c USN Malaria Control Unit Cactus. Here is a little color about the natives. I never saw any of the natives on the islands down there that ever had anything on their feet. They went barefoot and the soles of their feet were like leather. It was not uncommon for them to put out a campfire with their bare feet.

Most of the natives we saw and worked with were tattooed and that included the women. Many of them suffered from Yaws, and had open sores on their legs and bodies.

Yaws is a disease prevalent in very humid, tropical locations, generally close to the equator. The Solomon Islands fit that description to a tee. With an average rainfall exceeding 164 inches and just seven degrees longitude south of the equator, yaws is a very common affliction amongst its generally poor, native inhabitants.

Very similar to syphilis, but not sexually transmitted, yaws is caused by the spiral shaped bacterium classified as Treponema pertenue. It almost always occurs in the children first. Once the infection is introduced, a sore, or the mother yaw, appears after just three or four weeks of incubation. The sore appears on the skin as a raspberry-like growth and is highly infectious and very itchy. Children scratching the growth cause it to spread rapidly to other parts of the body. Yaws can be spread very easily to others by directly contacting the lesions or growths.

Yaws lesions can heal over time even without treatment, however, in most cases left untreated, yaws can be crippling; destroying skin, muscle, and bone tissue. Disfigurement is also a common side effect resulting in deformities of the nose, jaw, and legs.

Like syphilis, yaws can be readily cured with heavy doses of common penicillin and, therefore, is easily treated today.

Brian Taylor, PhD, Nottingham, UK, Onetime Government Entomologist Malaria Eradication Programme, Solomon Islands. An extremely effective anti-yaws campaign was conducted in the early 1960s, led by Dr. Jimmy MacGregor, later Director of Medical Services, and this eradicated the disease.

Robert Michel, PhM3/c USN Malaria Control Unit Cactus. Virtually all of the natives were forever chewing betel nut. They would put one in their mouth along with some leaves of some sort and some lime. The lime they made from coral. They would gather coral, put it in a fire to kill any organisms and then crush it into a powder. Each one of them carried with them a container made of bamboo, about an inch or two in diameter and about four to five inches in length. Into this they would place the lime powder. They also had a small thin stick, and when they were chewing the nut and leaf mixture, they would dip the stick into the lime and then into their mouths. The result was a very bright red spittle and jet black teeth. I think the betel nut was also some kind of a mild narcotic. I was going to try it once but got cold feet.

A little more detail might prove helpful here. When Bob describes the local natives on Florida chewing betel nut, he is describing an activity that has been more popular than the use of alcohol or tobacco has ever been in our society. A by-product of the betel nut palm, Arecu catechu, the seeds of the palm by that name have been cultivated all over Asia for hundreds of years. First mentioned as far back as 340 B.C. by Herodotus, the seeds of the betel palm—more commonly referred to as betel nuts—have been chewed for their mild stimulating effects and general feeling of well being.

The betel nut is a rather large item and is usually sliced into smaller, more manageable pieces. The slices of the betel nut are sprinkled with the burnt lime which Bob describes the natives extracting by fire from the local coral. The mixture is then placed in the mouth much like a hunk of chewing tobacco, and for several hours the mass of nut is chewed and sucked on without swallowing. What happens is this: the lime from the coral reacts with the betel nut, releasing a central nervous system stimulant called arecoline. Without the lime, chewing the betel nut alone will not release the arecoline and, therefore, the desired effect. The arecoline increases respiration and decreases the work load on the heart; in other words, a mild stimulant producing an overall good feeling for the user.

Often, users will moisten the leaves of the betel pepper and smear the lime on the leaves, then, add cloves, cinnamon, nutmeg or other spices to enhance the flavor of the experience. As Bob describes the natives on the Floridas he worked with, frequent, long term use of the betel nut mixture stains the mouth, teeth, and gums a bright red often so dark it appears black.

Chewing betel nut, however, should not be compared to the habit of chewing tobacco. Unlike tobacco, betel nut does not contain any of the harmful ingredients that tobacco delivers. In fact, betel nut is said to provide a number of medicinal benefits and is sometimes used in the treatment of stomach disorders, fevers, malaria, scabies, and even smallpox.

Robert Michel PhM3/c, USN Malaria Control Unit Cactus. I remember hearing the natives talking about God from time to time. I saw one of them carrying a Bible around. My guess was many of them had been converted to Christianity. Before the war, there were a number of missionaries on these islands and I am sure they converted a number of them, just as the missionaries did on Hawaii.

Historically, throughout the Solomon Islands pipe smoking was a revered habit. Originally picked up from early traders and sailors passing through the islands as early as the voyages of Captain Cook and other Europeans, it seemed the entire population always had a pipe at hand. Smoking was not limited to the male populace, but was embraced by women and children as well. Babies not yet old enough to walk but old enough to grip a pipe puffed away like adults.

Nggela family, 1943
Mother Smoking

Robert Michel, PhM3/c, USN Malaria Control Unit Cactus. Ah, the pipe smoking! Even the women and little girls smoked pipes and yes, little boys, too. The natives even grew their own tobacco. It was not uncommon to see a young woman holding her baby while smoking her pipe.

One of the chiefs was the father of one of my boys and one day he invited me to a party at their village. The village itself didn't seem to have too many huts in it for living in, but there were about 30 to 40 people around.

I went over there taking with me a rifle and some ammo that I gave to the chief as a gift. I also had with me a few blocks of TNT that we placed at the bottom of a lagoon. One of my boys named Andrew sat up in a tree on the edge of the lagoon to look down in the water to see when a school of fish would swim over the TNT. There I was sitting up against the trunk of the tree with the detonating battery in my hand, waiting for his signal. Pretty soon Andrew says, "Missa Bob, fish-e-come." As I got ready to fire the charge Andrew would say, "Noyet." Then a long pause, and again, "Noyet." And then, when the school was directly over the TNT Andrew said, "OK."

I set off the charge and all of a sudden there were people all over the place diving into the water to harvest the fish. They were piled up about twenty feet down on the bottom of the lagoon, two and three deep. All of us dove into the water and then down to the bottom where the fish were. Everyone had some kind of sack or T-shirt to stuff them into and go back up and empty them into the native boats that suddenly appeared. We must have harvested a thousand mullet.

Some of the people were Fijian troops who were apparently bivouacked nearby and arrived with Jeeps that they completely filled with the fish. I never even knew they were there until the TNT went off and then, when it did, a whole bunch of them came out of the bush and dove into the water along with the rest of us. I remember that they had at least two Jeeps, maybe three, which they filled to overflowing with the fish. That was the last I ever saw them. I don't even know where they were camped; it couldn't have been very far from where we caught the fish, otherwise, how would they have heard the explosion?

Marine Sergeant, Section Leader 75mm Pack Howitzers, I/3/11. A lot of the guys were doing that all the time. They would throw a charge in a lagoon or river and the explosion would stun the fish so they could jump in and pick them right up.

Brian Taylor, PhD, Nottingham, UK, Onetime Government Entomologist, Malaria Eradication Programme, Solomon Islands. On my first tour there in 1969, the islands were still littered with relics of the battles fought there. The locals were in the habit of using "lost" ammunition for fishing with explosives; sometimes hurting themselves in the process.

Mention of the Fijian troops stationed on the Floridas makes Bob's narrative all the more interesting considering the scores of books that have been written on the South Pacific Theater with nary a word devoted to any islander's participation in the regional events being played out. Once again, Bob's story forced me to expand my research into both the history of the Guadalcanal campaign as well as the local geography. In other words, dig a little deeper.

Any reference to Fiji leads one to believe that it is a single island located somewhere in the South Pacific. The Commonwealth of Fiji, however, is in actuality, a cluster of over three hundred islands strategically located smack dab in the middle of the vital shipping lanes connecting North and South America to Australia and New Zealand.

Originally referred to as the Cannibal Islands, Fiji, at the time of World War II, was a British colony, later gaining status as a full partner in the British Commonwealth in 1970. In 1987, Fiji declared herself an independent republic and was subsequently expelled from the Commonwealth. Today, Fiji has taken her rightful place as an independent nation in the world's community.

Though never directly attacked by the Japanese in their drive across the Pacific in World War II, the British colony of Fiji had a history of fierce loyalty to the Western Powers. Fijian's direct involvement in global conflict dated all the way back to the First World War when her chief, Ratu Sakuna, was highly decorated for his service with the French Foreign Legion. Fijian military units further aided the British throughout that war in non-combative roles in Europe.

At the outset of hostilities with Japan in 1941, Fiji was heavily occupied by Allied forces and their bases on Fijian territory became a necessary stopping point for the troops and supplies being fed into the battlefields of the Solomons and later, New Georgia. Hundreds of U.S. ships passed through her ports and thousands of tons of vital war materiel transited her harbors.

Fijian participation did not stop at opening her doors and her harbors to the Allies, though. Fiji contributed her young men to the fray just as mothers and fathers of the United States and Britain offered their children up to the war effort. Troops of the Fiji Infantry Regiment and the First Commando Fiji Guerillas fought side by side with American, Australian, and New Zealanders. They fought with distinction in the Solomon Islands and later in the New Georgia campaign, as well as in the fierce fighting on Bougainville. It should be noted that Fijian units also included native Solomon Islanders as well as residents of the independent Kingdom of Tonga. The native islanders were well equipped for life in the jungle and were said to have a night vision far superior to that of their western counterparts. Legend has it that many of the native troopers were able to spot Japanese snipers in the dead of night at distances up to one hundred yards.

Lt. Cmdr. Lewis P. May, USNR Ret., Assistant Supply Officer, USS President Hayes APA-20. We picked up Fijian troops I believe, in Suva, Fiji, and transported them to Guadalcanal. They were marvelous men, very large and cheerful.

They embarked on the President Hayes in the afternoon. I was the Assistant Supply Officer and my responsibility was to provision the ship. We had a lot of sides of beef to load and the Fijians provided the working party to load the large sides of beef into our refrigerated storerooms. They did the job with great gusto, but I do remember that each soldier deposited his side of beef at the entrance to the deep freeze compartment only. Apparently, the Fijians were very warm blooded and did not like the cold air inside the compartment.

I remember, too, that on that evening, a large crowd of local Fijians came to shipside and sang beautifully, singing goodbye to their beloved soldiers.

During the passage to Guadalcanal, I remember that they had huge appetites and delighted in our Navy chow.

Since we only touched at Tulagi on August 7, 1942, for the one and only time there, and since we never went into Purvis Port after that, I conclude that the Fijian soldiers we carried were delivered to Guadalcanal on one of our several stops there.

I am quite sure that the Fijians were infantry. It was also rumored that they could smell the enemy and were able to surround the enemy in the jungle.

Bill McLaughlin, PFC, 221st Field Artillery Battalion, Americal Division. Almost a year later, in 1944, we were sent out to Bougainville with the Fiji Battalion as field artillery (FA) spotters. After nearly a week out in the bush, during which time the Fijians impressed us with their fierceness, it was said their New Zealand and Aussie officers sent them off on a bayonet charge against enemy positions. The Japs bolted at the sight of them and the officers had to blow their whistles to make the Fijians leave off the attack and return.

We always had the ultimate respect for these people. For a group of guys who were brought up to disrespect blacks, as we were in those days, it showed us that all people with black skin were not the same as they had been stereotyped.

I'm sure it was the U.S. Army's method back then of always putting officers who were white over the blacks in any of our units. As officers, the blacks could only rise in rank to captain. White officers, and I have known some personally, couldn't wait to get out of a black unit. This hurt morale and led to people condemning the blacks as fighters—which wasn't at all true.

Robert Michel, PhM3/c, USN Malaria Control Cactus. Once while at the village, I could hear this terrible screaming, like the sound of a child in great pain. As we got closer, I realized that it was a pig screaming. One of the older members of the tribe had this long spear stuck in the pig. The pig was propped up against a tree. The man was having a casual conversation with another man and every once in a while he would look down at the pig and push his spear into another area trying to hit the heart and kill it. Every time he did that the pig would let out this terrible scream. He finally found the right spot and the pig died.

As it turned out, they were killing the pig in preparation for a feast. I never forgot that sound, though, and the casual way the natives regarded killing the animal. Later on, however, the pig was cooked and was delicious.

We had something else to eat that day, although I never did find out what it was. It was soft like dough and was formed into a loaf and placed into a stone oven and baked. When it was done, it had turned the color of purple and had an oily appearance. I don't remember what it tasted like; I often wondered if it was breadfruit.

All in all, it turned out to be quite a day.

Brian Taylor, PhD, Nottingham, UK, Onetime Government Entomologist Malaria Eradication Programme, Solomon Islands. This might have been the fermented banana common to the island of , or the taro, or cocoyam, grown in swampy land; but probably not breadfruit, which I do recall as commonly used in the Solomons.

Robert Michel, PhM3c, USN Malaria Control Unit Cactus. Another day Graves asked me if I wanted to go fishing. I said yes, but told him that we didn't have any fishing equipment. He said, "No worry," and he set it up for the following morning.

Most of the native boats that I came in contact with on the islands were dugout canoes. They were fashioned out of a single tree trunk, I think from the banyan tree that lined the shore and had huge roots. The trees could be found in the swampy areas abutting the shore of the channel between the Floridas.

For the most part, the dugouts were about ten to twelve feet long and about two feet in the beam. By my standards I would say they were pretty crudely made, but considering what tools they would have used, they were works of art. Can you imagine the number of man-hours required to make one of those?

So, the next morning we went down to the beach to a dugout canoe strewn with what appeared to be all kinds of trash in the bottom of it. A boy in front and a boy in back had paddles, with Graves and I in the middle. As we headed out to sea, Graves was building what turned out to be a kite, made from the trash which turned out to be banana palm leaves. He also had a large ball of cobwebs that the boys had collected in the bush the day before.

Graves was rolling this ball back and forth between his hands, making it into the shape of a snake, like you used to do with clay when you were a kid. He tied one end of a string to the middle of the snake, the other end to the tail of the kite. Yet another string was attached to the kite itself.

The boat turned around and headed toward shore, into the wind. Graves sent the kite up just so far as to allow the cobwebs to flick along the top of the water. Then WAP! Down came the kite. Graves pulled in the kite and the line attached to it. On the end of the line was a Gar fish. They have a jaw filled with teeth that got tangled up in the cobwebs and couldn't get loose. Graves pulled it into the boat, put its head in his mouth, and crushed its skull with his teeth. Then, the same thing all over again.

Now if that's not the darndest fish story you ever heard. Every word of it is true. We caught three or four fish, paddled home, gutted the fish, wrapped them in tei leaves, put them on the fire and ate some of the most delicious fish I have ever eaten.

The tei plant, by the way, abounds in the Pacific and its leaves are what the original grass skirts were made of. The leaves are also used extensively in cooking in the Solomons. Today I have a small tei plant in my greenhouse here in Carnegie that I hope will get big enough so that I can try cooking a fish the way we did on Little Florida.

Jutting out into the water from shore near my house was a short wooden boat dock, used to tie up small boats and our Higgins supply boat. One night, Chief Petty Officer Rountree, who was on Florida with me for a while, was out on the end of the dock with me, talking. We saw a large dark hulk move slowly through the water beneath us, right where we were standing. Rountree said it was some kind of fish and he was going to try to catch it.

The next day, he made a big fish hook from a steel rod. It was actually part of a canister used to carry 90mm shells. He used a file to sharpen the tip, then bent it like a hook, fastening a ¼" manila hemp line to it. That night we went out to the dock, baited the hook, and tied one end of the line to the dock for fear that if this huge fish took our bait and ran we wouldn't be able to hold him by hand. We lowered the hook into the water and waited. Sure enough, along came the fish and grabbed the bait. But he didn't run, he just sort of started wandering off. So with no fight we just hauled him up on the dock by hand. He looked like he weighed about a hundred pounds. Stuck in the roof of his mouth was what appeared to be a lobster claw. He couldn't get it out and it was decaying. Boy, did it stink! Rountree said we had caught a sea bass. Whatever it was, it was big. We cut off his head, cleaned that fish and cooked it at the Marine mess. All of us ate it and it was pretty good.

Islanders making grass skirts using leaves from tei plant

Every week or so a Higgins boat would come up the channel to our outpost with supplies, including drums of diesel fuel assigned to me. One day, I saw that one of the drums looked different from the others, so I opened it. It was alcohol…the good kind, 95% pure…the kind you can drink! There were twenty-five Marines and one sailor bombed out of their minds for weeks on alcohol, warm water and lime juice.

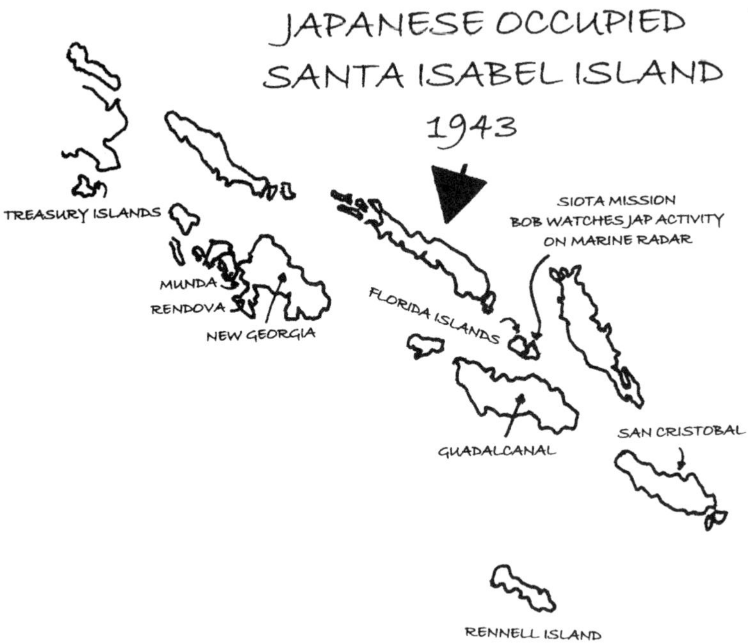

One night I was in the radar shack and we saw unfriendly blips on the scope that looked like the whole Japanese army was out there. We radioed the Canal to report what we saw. They, however, accused us of being drunk; which we were. A few days later a G2 report came out that the Japs had evacuated Santa Isabel Island, 50,000 troops. At least that was the story we got. Drunk or not, the screen was solid with blips.

Santa Isabel Island, located in the middle Solomon Island chain, can be located on a map by looking just to the northwest of

Guadalcanal. The island is a mere eighty-three miles long and approximately fifty miles in width at its widest point. Santa Isabel is made up of a single chain of volcanic mountains which reach a maximum height of 4,000 feet. Dating back to its discovery by the Spanish in 1568, Santa Isabel Island, with five fast moving, fresh water rivers cutting through the island, was a favored watering stop for ships plying their trade in the area. As with many other islands in the chain, though, Santa Isabel was of little strategic value other than for a seaplane base built by the Japanese at Rekata Bay in the spring of 1942 as they advanced further south and east toward the New Hebrides and New Caledonia.

Early Spanish explorers and traders, originally smitten by the friendly natives who inhabited the island, soon became disgusted when they learned the islanders, who worshipped snakes, lizards, and crocodiles, were also inveterate cannibals. In later years, however, the island had become a Christian landmark in the area, with a strong missionary presence established there. The last reported instance of cannibalism was reported just forty-six years ago, in 1954.

After Guadalcanal was declared secured by U.S. Army brass on February 8, 1943, United States Navy and Army planners were yet unclear where the next Japanese amphibious landing might take place. Guadalcanal and surrounding islands were located in the middle Solomons and were even then susceptible to an annoying pressure by strong Japanese forces located further up the Solomon's chain. Just as U.S. planners were in doubt over Japanese intentions, so too was the Japanese high command in Truk and Rabaul shaking their heads. They were unsure as to which island would be the next stepping stone in the imminent U.S. advance.

At the time, Navy brass had their collective eye on two locations: the New Georgia group, with its potential for advanced air bases on the island of Rendova (more specifically at Munda Point on the southwest tip of New Georgia), or Santa Isabel Island with its developed seaplane base at Rekata Bay. There was no debate, however, that removing the Japanese from their stronghold at Rabaul would be their number one objective. Santa Isabel, therefore, which lay to the northeast of Rabaul, was not the most direct route to achieving their goal. Elevating an attack on New Georgia, either at Munda Point or across the harbor at Rendova, represented the most obvious location for Allied attention. The Japanese garrison at Santa Isabel Island, therefore, became one of the first of many over the next two years that would be bypassed in the developing U.S. strategy of island hopping.

Back to Bob's story. When he described the radar activity he and the Marines saw that night on Pile at the radar installation, there were a mere 3,000 Japanese defenders on the island of Santa Isabel. More than likely, Bob and the Marines did, in fact, witness a great deal of activity up on Santa Isabel, but an evacuation of 50,000 Japanese troops, it was not. What Bob and the partying Marines more than likely witnessed was the Allied attack which took Rekata Bay from the Japanese.

The Japanese garrison on Santa Isabel would remain in place almost to the end of 1943 before it was eventually captured by U.S. forces.

Brian Taylor, PhD, Nottingham, UK, Onetime Government Entomologist Malaria Eradication Programme, Solomon Islands. For some years after the war, there was a story of a Japanese soldier, or soldiers, still living in the forest on San Jorge islands, off southeast Isabel. In 1971, or thereabouts, a group of Japanese searching for such lost men came to the Solomons and spent a frustrating and fruitless time on San Jorge. After they gave up the quest, the local Government Information Officer, Bill Bennett (I think that's who it was) admitted that he had led a party in the late-war period that had found and killed the apocryphal survivors. He, however, had derived great enjoyment from the discomfort of his old enemy in their fruitless efforts.

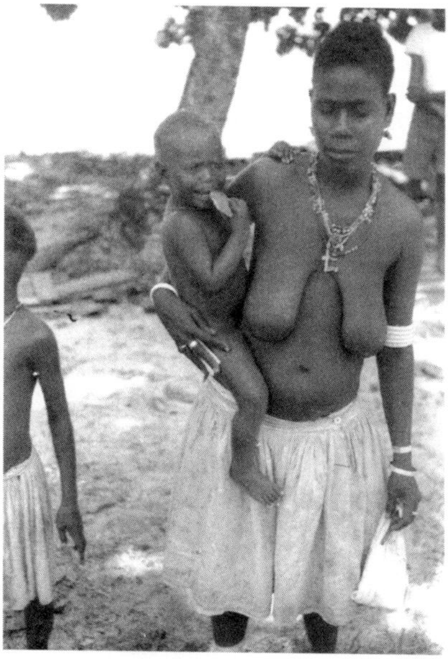

Mother and child
Gila Kiki, Solomon Islands, 1943

Robert Michel PhM3/c, USN Malaria Control Unit Cactus. I had a lot of free time on Florida so I wrote to the International Correspondence School and enrolled in their high school program. As you may remember, I never finished high school. I joined the Navy having just completed my junior year. The correspondence school would send me a lesson and a test sheet. I would read the lesson and fill out the test and send it back to them. When they received my test, they would send me another. That must have gone on for five or six months—until I went to New Zealand for R & R—and that was the end of the International Correspondence School.

Back to the New Hebrides…

Dave Minton, PhM3/c, USN Malaria Control Unit Cactus. I think it was probably around the first week in January, 1943, when

I was transferred back to Espiritu Santo. I was assigned malaria control duties around the airfield there and also covered areas wherever they had troops stationed. I remember a French planter that lived in a house near there much like the house in "South Pacific"—you know, the play that came out a few years later.

We had an armored vehicle that we used to get around in and we would fill it up with our tanks and oil and head out to find open water to spray.

We needed to get along with the natives around there because we had to have them work with us. So, we had to know a little of the pidgin to communicate. Frequently we came in contact with natives who lived in the coastal areas of the island and sometimes had them work for us. I remember they hated the natives who lived in the interior and vice versa. In fact, we were told of a tribe that lived farther back in the bush that we were to leave alone; we were not to go into the area they occupied. So we went just so far and never any farther, just to stay away from those natives.

The inland natives were dark skinned and very black. They derived their subsistence from wild pigs native to that area. In fact, they lived with them. They actually captured the wild pigs and in captivity they knocked the pigs' front teeth out so that their tusks would grow into complete circles. Solid ivory—the circular tusks were quite valuable and worn by natives as jewelry and also used as currency amongst themselves and other tribes.

I collected as many of these as I could and saved them in a shoebox. When I finally returned to the States, I took that shoebox back with me. I gave one of these ivory bracelets to every girl I knew, thinking they were so valuable. I am sure the girls all threw them in the garbage as soon as I was out of sight. I still have three or four of them, however, hanging on the wall in my house after all these years. One of them I gave to Bob's wife, Marj, when I was up at their house on my way to Connecticut a few years ago.

It was while I was stationed at Espiritu that our malaria control unit was attached to the Army—it was the 43rd Infantry Division, the 103rd Regiment. These were all guys from Connecticut and Vermont who, if you think about it, should have been my friends, so to speak, since I'm from Connecticut. These were guys who had been in the National Guard for years and they were old—not 18- or 19-year-old kids like most of the rest of us were. Most of them were at least ten years older than I was. They just weren't lean and mean like a lot of the younger guys were and lacked that hardness I guess you need in combat to survive These were guys that had come in on the Coolidge, which didn't help them at all. I was to serve with these guys later on Rendova and at Munda.

Thomas Gross, U.S. Army 97th Field Artillery Battalion, Americal Division. I finally left Guadalcanal on May 8, 1943, on the Vitis Bering, a Liberty ship on its maiden voyage from the States. I had stopped taking my prior to leaving the island. The trip from Guadalcanal to Noumea's outer harbor took a whole week. Then, we were another week waiting in the outer harbor; another week in the inner harbor, and finally, a week tied up at the dock. Right around the second week into the trip, I started to feel sick. I didn't want to be put into the hospital in Noumea, so I started taking the pills again.

As it turned out, the Army pulled us off the Vitis Bering and put us on a ship gong to the States. While on that ship, I stopped taking the pills again and got sick as a dog. They put me in the sick bay on the top bunk. I called the medic for help on the second day in there. Now, to the left of me, was this guy with his eyes all bandaged up. When the medic came, I asked him what was wrong with the guy. The medic told me he was blind. That stopped me from ever asking for help again on that trip.

When I got back to the States, I was transferred to OCS (Officer Candidate School) in Fort Sill, Oklahoma. I was in the first four-month class, Class #41. Well, as it turned out, I would go to school for a month and start to get the chills again, so I went back to the hospital for another month. Back to school for a month; then back to the hospital for a month. In OCS, we had several sections of training, such as gunnery, etc. If I became sick toward the end of a section, I would take a lot of quinine so that I could finish that section.

I remember one night I was so sick my hut mates called the medics to come and get me. I wanted to stay a couple of days more, and when they tried to put me in the ambulance, I put up my feet and wouldn't go in. They told me I would get thrown out of OCS if I didn't get in the ambulance. I guess I was so hopped up on the quinine or so sick I said, "The hell with OCS—throw me out!"

So, finally, after starting OCS in Class #41, I finally got my 2nd Lieutenant Bars with Class #97. I still kept getting sessions of malaria for the next five or six years. I had to sign a release on malaria just so I could get my commission.

No problem with the malaria now, though I still take quinine sulfate. Six hundred milligrams, every night, for the cramps I get playing tennis or golf.

Chapter 10

The Central Solomons

June 1943-May 1944

Dave Minton PhM3/c, USN Malaria Control Unit Cactus. At the end of June 1943, I was still with Malaria Control and there were a few other guys there in the unit with me. We left Espiritu in a convoy of ships and I was on an LST, as were most of us. We first went to the Russell Islands where we stopped off and picked up the 103rd Infantry who were already there. From the Russell Islands we went straight to Rendova.

I'll never forget that day we were going in to the beach. We saw about a hundred planes heading right for us, coming in about fifty or sixty feet off the water. We thought we were all going to be killed right then. We all just hit the deck in the bottom of that boat. I'll never forget what it was like trying to dig a hole in the two-inch thick steel deck with my fingernails. Thankfully, they just flew over us. It turned out they were our planes and we didn't get killed that day.

We actually landed on Rendova about three or four days after the first landings, so it was secured when we got in there. It looked like there had been a lot of bombing there because debris was lying all over the place—equipment, crates, blown up trees, and the like—just piles of it everywhere. And the stink, that I'll never forget. The place just smelled putrid—it was horrible.

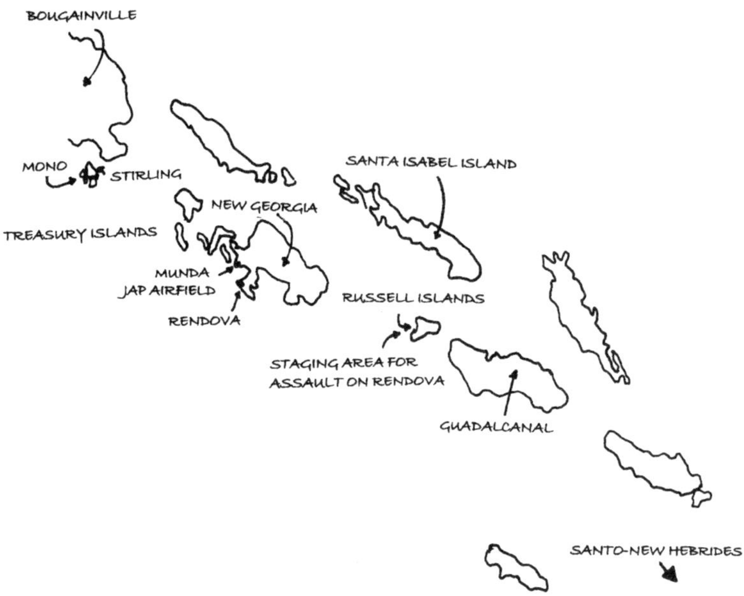

ARMY LANDS AT
RENDOVA-1943

When I got there, the Army was firing their 155s across the channel at the Japanese on Munda. Those guns were set up right behind us and it was really something to have those things going off right next to you all night long. Whuuump. Whuuump. That was terrible. The shells went right over our foxholes.

While I was on Rendova, I had a couple of my teeth pulled by one of the Army doctors. They also were very short of medical personnel on the island, so they had some of us from Malaria Control help out. We were all Corpsmen, you know, so they had us assigned to the hospital for a while.

When the Army crossed the channel and attacked the airfield at Munda they were bringing all the wounded back over to the hospital on Rendova for treatment. The place looked just like the hospital in the movie "MASH." They had fifty-five gallon drums set up with stretchers lying across them with the wounded just lying on top of them. I'll never forget they were bringing in these poor guys who had been hit real close up in the fighting and their skin was all blackened from the shells and explosions. Their skin was just horribly burned. I will never forget cutting the black flesh off of them using a scissors and then putting sulfathiazole on the open wounds. They were then evacuated off the island for more treatment.

Today, I can't stand the sight of blood—I can't even watch an operation on TV. If I see one come on, I just turn it off. Back then, though, it didn't seem to bother me—all the blood and the wounds. I didn't mind it—I guess it was just part of the job or something. I didn't really notice it that much, I guess.

B25 Just short of the runway
Munda Fighter Strip, 1943

I was still with Malaria Control a few weeks later when we went into Munda after the beach was taken there. That was a real mess over there, too, and the Army was in a lot of trouble. It was a real heartrending experience.

The 43rd Infantry Division later reported experiencing a great deal of difficulty in overcoming Japanese resistance at the Munda airfield at the end of June 1943. A large number of the troops were identified as suffering from unusual medical problems. Only ninety men had been killed in action up to July 17, but more than 1,000 men had been taken out of the action by these medical problems. It was later reported that the bulk of these casualties were not caused by wounds or infectious disease but by mental disturbances. As many as 150 men a day were leaving the lines and heading for the beach with what was described as war neurosis.

Dave Minton, PhM3/c USN Malaria Control Unit Cactus. There was no close-in fighting going on when we got to Munda—the airfield had been secured and there was no more ground fighting going on. We were stationed on the top of a hill that gave us a real good view of everything around us, including the airfield. Even though the ground fighting was over, the Japs still came in and bombed us every night.

I don't remember any of the guys' names I was with except a guy named Schwintaurus—everybody called him "Ski." He was in our malaria control unit. We were getting bombed one night and Ski jumped into our foxhole to take cover and injured his leg or twisted it or something. He wasn't hit at all from the bombs, he just hurt himself jumping into that hole. He started to complain that he was in a lot of pain after that and everybody kept telling him he was faking it. He had a doctor check it out and the doctor told him he couldn't find anything wrong, just get back to work. Ski kept complaining about the pain, though, until he was finally sent back to a hospital on Guadalcanal. The story we heard was a doctor came in to check him out and Ski told him he thought he had an embolism or something in his leg. The doctor just lifted his leg straight up, telling him there was nothing wrong with him. With that, Ski said, "You son of a bitch—you killed me." And Ski died right then. That's the scuttlebutt we got back, anyway. Of all the guys I was with, I still remember Ski.

On Munda, we went around and sprayed oil on every piece of water we could find, and there was a lot of it. We covered everything with oil, to control malaria, of course—everything, that is, except one place. Off toward the Bougainville side of the island, I think that would be on the west side of the island, there was the most beautiful freshwater pond I've ever seen. It was about 100 feet deep but was so clear it looked like it was just ten feet deep. When you threw a rock in that pond, I swear it would take the rock five minutes to get all the way to the bottom and you could watch it the whole way. The water was icy cold when you jumped into it. It was really beautiful.

Munda Fighter Strip
View to the north 1943

While I was at Munda, I met Pappy Boyington. He was the best fighter pilot we ever had. He got most of his kills flying out of that airfield. The way he did it was when they flew into a bunch of Zeros and all the planes were mixing it up in a dogfight, he would stay off to the side and conserve his fuel. When the planes that were left started to head off back to their bases, he would pounce on any Zeros that were left. They had loudspeakers set up around the airfield so we knew when he was coming in. We could hear the ground controllers talking to the pilots and would watch them come in as we listened. When all the planes had landed, here would come Pappy towards the runway all alone. When he got to the end of the runway and he was real low he'd do a victory roll. Then another. Then a third. Then a fourth. Four planes he had shot down—he was really something.

New Zealand and R&R

Robert Michel, PhM3/c USN Malaria Control Unit Cactus. My next stop was R & R at a hospital in Auckland, New Zealand, where once again Dave and I were together for a month. Afterwards, we were sent back up into the islands.

When Dave and I arrived at the hospital we spent the first couple of days eating real ice cream and fresh milk and good tasting food and sleeping. Then we got a leave, and I think it was Dave's idea (how he heard about it, I don't know) to take a train to Rotorua, where the Maori's lived at that time. Rotorua to me was one of the most fantastic places I had ever been to. That was our first introduction to Polynesia and we fell in love.

Anyway, we stayed at the Lake Hotel, or Lake House, that was situated, naturally, next to a lake. It was a lake of weak sulfuric acid. You see, Rotorua is a thermal region. I also believe the Maoris who lived there, did so for free. However, because of its thermal nature, small steam vents were all over the place and the Maoris used them to cook on. Each of the Maori families had their own vent. As a result of the steam coming up out of the ground, the vent was surrounded by a small area of mud. The mud also contained sulfuric acid and the Maori women and young girls would spread this mud on their arms and legs for a short while and then wash it off. It left them with the most beautiful, clear skin you ever saw.

In the hotel where we stayed, the rooms had very large bathtubs and the hot water was supplied by a subterranean source. When you turned on the hot water, out came this sulfur smelling hot water. It was fantastic. The lake was supplied by the same water source. On top of the water of the lake, and in the bathtub, would be a sort of crud that looked just awful. The appearance of the water, though, didn't stop people from using it. It also smelled awful, but it was the greatest soak I have ever experienced.

The first morning there, Dave and I ordered breakfast in bed and this beautiful Maori girl brought us fish, scrambled eggs, toast, and tea. It was the first time in my life I ever had fish for breakfast and I loved it. Later, when I got into the travel business and moved my family to Hawaii, my associates and I would go to the Ilikai Hotel for breakfast and have broiled kippered herring, an omelet, tea, and toast for breakfast. It's a great breakfast.

Dave Minton, PhM3/c USN Malaria Control Unit Cactus. I don't know how Bob and I got that R & R on New Zealand. At the time I never knew how or why things happened. If someone higher up was pulling strings, I never knew, but it was entirely possible. I just remember my name would be called and I would go. I remember well the good times we had though, especially that first morning when they brought us breakfast in bed—tea, eggs, and even fish. Fish, for God's sake. Bob sure does remember that right!

Bob Michel, PhM3/c USN Malaria Control Unit Cactus. While at Rotorua there was a party at someone's house every night. Dave and I would go to the hotel bar that only opened a couple hours each day and buy two burlap bags filled with quart bottles of Weidemeier beer. By law, you were not allowed to carry a bottle on the street unless it was in a bag or concealed in some way.

Anyway, down the street would go these two swabbies bent over with the weight of those sacks of beer. Then the party would begin and go into the early morning hours.

We made wonderful friends during our short stay, and when we were leaving, a whole group of them sang us a song that at the time was unheard of except in New Zealand. It was the Maori Farewell Song that was later recorded and became a big hit after the war. The first line you may remember: "Now is the hour for me to say goodbye, soon you'll be sailing far across the sea...." I told Dave we ought to copyright the damn thing. We didn't; someone else did and made a bundle.

Dave Minton, PhM3/c USN Malaria Control Unit Cactus. While Bob and I were at the hospital in New Zealand, we ran into a Navy nurse by the name of Mary who was going out with Boyington at the time. She was his girlfriend and they were dating. She worked at the hospital there and Bob and I used to run into her in the halls and sing to her—'Mary...Mary...." We got a big kick out of that.

Bob Michel, PhM3/c USN Malaria Control Unit Cactus. While we were at the hospital in New Zealand, Dave and I would go into Auckland on liberty. One night, while we were there, we met a couple girls we liked. So, every time we went on liberty we made sure to date these girls, usually going to a local cinema. There wasn't much else to do.

One night, Dave and I invited the girls to go out to dinner with us. We took them to the fanciest hotel in town and had one hell of a fancy dinner with white-gloved waiters attending to our every whim. The people in New Zealand did not have the kind of spending money the Yanks had, and these girls had never even dreamed of going to a place like that. After the dinner was over, and remembering the lessons of their mothers, they proceeded to scrape all the dishes on the table, just as they would at home. The waiters were from the old school and didn't bat an eye. They just smiled and removed the dishes from the table.

Of the many things Dave and I remember about Auckland, that incident stands out the most. They were sweet, young, naïve girls who had never been exposed to some of the manners associated with eating in a high class restaurant or being served by waiters in white gloves. Things were much simpler there, almost like it was in the States during the Depression. Nobody owned land there as I remember, and no one had an electric refrigerator, only an ice box.

Another thing about the girls. If we got the last tram from Auckland and got them home late, it meant Dave and I couldn't get back to the base. The trams stopped running after ten o'clock at night and we had to get the girls home before 10. So, if we got the last tram from Auckland, we quite often ended up sleeping at their house and going back to the base the next morning.

It was a fairly common occurrence for guys on liberty to spend the night at their date's house because they missed the last tram. I heard a story about one sailor who brought his date home and was sitting around talking with the family when the mother said, "Why don't you and Lois go up to her room where you can be alone?" They did, and ended up in bed and fell asleep. The next morning, when everyone had gone to church and the sailor was sitting in the living room, the mother came in and said she wanted to have a talk with him. She informed him that just about all the men in New Zealand had gone off to war fighting for England. The women who were left at home were truly concerned about the propagation of their race as a result of so many of their men being killed. The mother told him that if Lois were to become pregnant, they would not force him to marry her but would welcome the newborn into their own family.

Now I have to jump ten years to 1952. I was working at a TV station, W H E N in Syracuse, New York, and was the host of a lunchtime fashion show at a local hotel. The fashion show was televised. One day, after I had been doing the show for about six months, the woman who described the dresses to the audience, the hotel coordinator, and myself were sitting down to lunch after the show. A girl came up behind me and asked, "Poden me, but were you in the Nyvy?" I turned around and said, "Yes." She then asked, "Were you ever in New Zealand?" My God, I thought, has my past finally caught up with me? I told her, "Yes." Her next question then sent chills down my spine. "Were you ever in Auckland?" She then asked me if I had known a Lois Cocker. When I replied yes to that question and recovered some semblance of composure, she informed me that she was Lois's little sister and that she had married a Yank and was now living near Syracuse, New York. She had seen me on the show and had recognized me. She had written back and forth to her sister about me. Finally, she got up enough courage to come to the show and talk to me. Talk about small worlds.

Another New Year—The Northern Solomons, 1944

During World War II, shore based duty for Naval Hospital Corpsmen and Pharmacist Mates scattered them in hospitals and dispensaries throughout the United States and all over the world. In the South Pacific, as the advancing U.S. military captured more and more islands from the Japanese, the Navy established Advanced Bases and Hospitals as an integral link to evacuating casualties from the battlefields at Espiritu Santo, Guam, Kerama Retto, Kwajalein, Leyte, Majuro, Manus, Tongatabu, and Ulithi. Fresh from the brief respite afforded him on New Zealand, Bob Michel was sent to one such Advanced Base Hospital designated only with the numeric code name, Navy 811.

Once Guadalcanal was secured, Vice Admiral Halsey's forces in the South Pacific, supported by the Thirteenth Air Force—nicknamed the Jungle Air Force—initiated an offensive campaign in the Northern Solomon Islands with the objective of retaking the Japanese stronghold at Rabaul. Continued air superiority lay at the heart of Allied strategy. To effect this goal, airfields were established almost before the smoke of ground combat had cleared. As early as February 1943, the Russell Islands were attacked with the sole purpose of establishing a forward airfield there to press the attack on New Georgia. At the end of June 1943, Rendova was assaulted, followed by Munda in August, and Kolombangara shortly thereafter. The Treasury Islands were next on the list.

On October 26, 1943, troops and supplies were loaded up at Guadalcanal, Vella Lavella, and Rendova, departing for the initial assault on the Treasury Islands. The Treasury Islands were just off the south coast of Bougainville and a mere sixteen miles from the Japanese held airbase at Shortland Island. Five groups of ships timed their departure to arrive simultaneously at Blanche Harbor located between Mono, the larger of the two, and Stirling Islands on October 27. Timing was critical to the success of this operation because of the close proximity of available Japanese reinforcements stationed on Shortland. For the operation to be successful, it was necessary for members of New Zealand's 8th Brigade, who went ashore in the initial assault, to quickly establish their beachhead and set up defenses.

As it turned out, only about 225 Japanese Special Naval Landing Force troops were holding Mono Island. Heavy naval gunfire forced the Japanese defenders to take cover in the hills, and by noon all resistance to the landing by the New Zealanders had ceased. Mono Island was firmly in Allied control. By the November 12, the New Zealanders had finally killed or captured every member of the Japanese occupying force that had fled to the hills on the day of the assault.

Stirling Island, just across Blanche Harbor and not occupied by the enemy, was quickly taken by 2,500 Allied troops, most of whom were also New Zealanders of the 8th Brigade.

From November 1, 1943, through January 15, 1944, the Allies continued to pour troops and equipment into tiny Mono and Stirling Islands. A boat pool was constructed, as well as an Advanced Naval Base (811), and radar was put into operation in an effort to support the main Allied operations against the Japanese at Empress Augusta Bay on Bougainville.

It was the United States SeaBees who constructed a 5,600 foot long airstrip on Stirling that started fighter operations Christmas Day, 1943. Bob Michel, fresh off his New Zealand R & R, was sent to Advanced Base 811 on Stirling Island to take up malaria control duties in the area right next to the airfield.

Once the airfield on Stirling Island was in full operation, it didn't take but another four weeks for the Thirteenth Air Force to transfer the headquarters of their B24 equipped 42nd Bomber Group from Guadalcanal to the new base. Starting on January 20, 1944, the 42nd, 70th, and 75th Bomber Squadrons were in operation on the island, launching raids from the newly completed airfield against Japanese naval and air bases at Rabaul in New Britain. Just five days later on January 25, a third squadron was added to boost the already powerful air arsenal on the island with the arrival of the 100th Bomber Squadron. Priority one in reducing Japanese bases by air was to keep the planes flying. One sure way to do that was to keep the pilots and ground crews of those planes healthy. The 42nd Bomber Group could ill afford an 80% malarial infection rate of its fliers. As the combatants advanced, therefore, so did malaria control expert Bob Michel.

By the time Bob Michel was transferred to Navy Advanced Base 811 on Stirling, he had been in the South Pacific for eighteen months; with six of those months on Guadalcanal living under the most horrific conditions. Every day brought bombings, shellings, and often, even worse, barrages from Japanese naval ships lobbing tons of explosives into the Marine perimeter. Couple these unceasing traumatic events with over a half century of elapsed time and it is little wonder that one's memory often dims and events run together. Such was the case when Bob and I turned our attention to his six month tour of duty on Stirling Island. Admittedly, Bob felt somewhat at a loss; frustration and sometimes even anger started to surface when I pressed too hard. In Bob's words, "Dennis, for the life of me, I only have a couple of memories of the place. I couldn't describe a blade of grass over there let alone any significant buildings or incidents." Suffice it to say, we do know that Bob's duties continued in malaria control, this time at an airbase caught up in the larger Cartwheel Campaign to retake Rabaul. Here's what Bob had to say.

Robert Michel PhM3/c, USN Malaria Control Unit Cactus. When I left New Zealand after my thirty day R & R, I was sent back up into the islands, this time to a place that was called something like Stirling. They had an airstrip there that I remember bombers flew in and out of. If I recall right, they were B24s.

I remember sitting out by the airfield one day, don't ask me why or what I was doing. I just remember sitting out there. I recall one of the bombers that came in that day had a problem with its landing gear or something. Anyway, when it came in, it slid down the runway on its belly. Those planes had a ball turret on the bottom, and when the plane slid in, it completely crushed the turret underneath the plane. I will never forget, the guys out there just about had to use a spoon to scrape the body of the poor gunner out of the turret. There was just nothing left of him.

Beyond that, I couldn't tell you a thing about the place. The next thing I really remember is getting back to Noumea, New Caledonia, and getting ready to ship out back to the States.

Not only was time starting to take its toll on Bob, life in the South Pacific was tough for many of the soldiers, sailors, and airmen who had by now spent well over a year and a half in the islands and were faced with the unknown prospect of just how long they might still be there before the war ended. South Pacific duty represented long periods of idleness followed by short periods of the most intense, brutal combat ever imagined. It was the boredom of the constant waiting, never knowing, and strange environs that finally started to cause some strange behavior to be exhibited by some of the men.

Robert Otto, BM3/c USN, 20mm Gunner, USS Grayson. Didn't you ever hear the story about the guys at Noumea who decided they were going home? No? Well, here's what they say happened. Some Navy guys got together and decided they had had enough and were going to steal a boat and go back home to the States. So they did. They got their hands on one of those tank lighters; you know what those are. They had it all stocked up with water and provisions, and one night they just took her out. They got well outside the harbor at Noumea before one of our patrol boats out there spotted them. They sent some boats out there to pick them up and bring them back in. I guess they put those guys in the brig. Can you imagine, that?

Dave Minton, PhM3/c, USN Malaria Control Unit Cactus. After Bob and I left New Zealand, we were both sent back up into the islands. I don't have any idea where they sent Bob and I didn't see him again until we hooked up on New Caledonia right before we shipped back to the States.

I was sent over to the Floridas for malaria control work over there. I was on the big island, Nggela Sule, and had my own house up on a hill right by a Navy radio transmitter tower. I remember I would wake up in the morning and I could look right out of the house from my bed and right down onto all the ships and boats in Purvis Bay. It was a beautiful place and while I was there, I had the best time I ever had the whole time I was in the Navy.

I had five houseboys up there that worked for me and took care of everything. At night I could even go down to the Navy Base in the harbor and watch movies. They had a big outdoor screen and stage set up there and we had a great time. I could eat at the Navy mess right there by the transmitter whenever I wanted to, and for

Dave's house overlooking Purvis Bay
Ngella Sule Florida Islands, 1943

the most part nobody ever bothered me up there. It was really a beautiful place.

When Bob and Dave went back up into the islands after their short, but well deserved rest in Auckland, New Zealand, malaria control efforts by all branches of the military in the South Pacific were well established. The techniques developed earlier in the war by units such as MCU Cactus and later, Army units, were being put into play theater wide. Effects of their work and the development of more sophisticated equipment such as the bug bomb, protective netting and clothing were paying off in drastically reduced levels of malarial infection among troops throughout the South Pacific.

View across Purvis Bay towards Tulagi
Ngella Sule Florida Islands, 1943

William K. Schultz, CM3/c USN Malaria Control CBM538 SeaBees. I joined the Navy SeaBees, Construction Battalions, and went through Boot Camp at Camp Perry, Virginia. We had Marine drill instructors during basic training and afterwards, we were formed into the 103rd Construction Battalion which was composed of four companies. We were then shipped to Quonset Pointe, Rhode Island, for our advanced training where we spent about four months. We were next shipped to Gulfport, Mississippi, where we continued to train until shortly before we were to ship out for overseas. At this time, we were informed that our Battalion, the 103rd, was to be decommissioned. We were split into four CBMUs: CBMU 536, 537, 538, and 539. CBMU 536 and 537 were shipped out about two weeks before us and no one had any idea where we were going.

On December 21, 1943, at about 7 P.M., we were given a total of thirty minutes to pack all our gear and assemble. We marched from our base to the docks at Gulfport, Mississippi. I remember it was quite a distance, but I'm just not sure now how far it was. Unknown to us, the ship had already been loaded with all of our equipment such as trucks, caterpillar tractors, road graders, and fuel.

When we shipped out overseas to the South Pacific, we had not been told where we were going and, even if we had known, it would not have meant a thing to us.

The ship we were to sail overseas on was the SS Azalea City. It was a brand new ship making its first voyage overseas, though it was not designed to haul troops. Metal frames had been installed four high with about 18 inches depth between them. The aisles were so narrow that only one person could move at a time through them. There were a total of about 1,400 of us aboard. Five hundred of us were with two units, CBMU 536 and 539. The rest of the men were made up of what was called a "casual draft." These were men that for one reason or another had missed their outfit's sailing, or had gone AWOL, or had been in military prisons. These men were apportioned out as replacements wherever they were needed. One of these men from the casual draft was originally from our outfit and had gone AWOL. He was later given a transfer back into our unit.

The Azalea City was not equipped well enough to feed us all, so we were only fed twice a day. We were issued different colored chow cards and your time to eat depended on the color card you had. The times for chow for the different colored cards were changed each week to even things out, since feeding the troops went on twenty-four hours a day or until everyone got their two meals that day. Mostly the food was terrible, and since we had to eat out of our mess kits, everything was thrown together and was not very nice to look at. After a few weeks of this, everyone was very hungry almost all of the time and we would steal any food we could get our hands on.

About this time, the man who had been AWOL and transferred back to us stole a loaf of bread that had just been baked. He ate the whole loaf, and then the dough swelled in his stomach and killed him. He was one of the oldest men we had in our outfit—I would guess he was at least forty-five years old. Anyway, since we were only about a day away from Bora Bora, it was decided to bury him there rather than bury him at sea.

When we arrived, I thought this was one of the most beautiful places I had ever seen. We dropped anchor and the natives came out to the ship by the hundreds in their outrigger canoes. It was like the Hollywood version of a South Seas island in that the women were all nice looking and most of the men looked just like Tarzan. The men stood up in their canoes and threw coconuts, with their husks on, up to us; which had to be at least 35 feet or more above the water. In return, we would throw them packages of cigarettes and coins.

We then lowered two boats and the body of the man who had died was taken ashore and buried. To the best of my knowledge, there is no U.S. military cemetery on Bora Bora. Several weeks later, another one of the men from the casual draft died as well. I am not sure what he died from but he was buried at sea. He was sewn into a canvas bag, a service was held on deck, and he was dropped into the ocean.

When the burial party came back aboard, they told us that CBMU 536 and 537 had landed on Bora Bora just a few days ahead of us. The chief there had a hut built for each of the men and had given them women to clean and do their laundry. Of course, we were all jealous when we heard this. After the war, when I would tell people this, they would think I was lying. Sometime in the late '40s or early '50s, however, **Readers Digest** did a story about the King of Bora Bora. The story went something like this. In the '20s, Hollywood filmmakers had come to the island to make the original movie, "Mutiny on the Bounty." This man had been an extra they had used in the film. It was said that American women liked him so much that they brought him to Hollywood, where he later made many pictures for which he was paid very little money. He wanted to go home very badly, so the filmmakers finally started to give him more and more money just to keep him in Hollywood. He became one of the most famous stars of the silent pictures. He finally returned to Bora Bora and became the chief there. He was still the chief at the time our ship stopped there and the other SeaBee units landed there. He was able to speak English very well and really liked Americans. This is why we were all treated so well at Bora Bora.

I have been thinking about going back, maybe this year, or perhaps we will wait 'til next winter. I have always thought that if I did make it back there, I would try to find the man's grave who we buried there. If I ever do, I will let you know.

We arrived after a six week voyage, leaving from Gulfport, Mississippi, and going through the Panama Canal at Espiritu Santo, New Hebrides, in February 1944. Interestingly enough, the Japanese had just stopped sporadically bombing the New Hebrides shortly before my unit arrived there. I don't think Santo was as great a place as Bora Bora; I would rate it an 8 and the beaches there a 10.

The natives at Espiritu were not a good looking people compared with the people of Bora Bora. There were a large number of Tonkinese there that the French plantation owners used as slave labor; at least that was my impression. Within thirty minutes of arriving, we heard screaming coming from a yard outside of a house. A Frenchman was beating a Tonkinese woman with a club. One of our men jumped across the fence to stop him but had a pistol pointed at him and, of course, that ended that. The Frenchman must have complained to our officers, since the next day we were told not to interfere.

There was about one Tonkinese woman for every seven men on the islands there. I do not know what the sharing arrangements were, but they were all pregnant or had just had a baby. Often we would see a woman carry two babies in slings, one on each hip and have three or more walking beside her all aged about a year apart. They all chewed betel nut so their teeth were very black, otherwise they actually would have been quite nice looking. As you know, this was where the character Bloody Mary originated.

William K. Schultz CM3/c, USN Malaria Control CMB538 SeaBees. Within hours of our arrival, a request for a volunteer for a job was made, but we were not told what that job was to be. As it turned out, I was the only one to volunteer.

On the New Hebrides, there was a unit called Island Malaria Control. Each unit had to provide their own support for covering an area of at least a mile from the perimeter of their camp. In my case, this consisted of jungle on three sides and beach on the other.

I was sent to classes which I think were given by the Army. These lasted for about a week and were mostly on teaching me the larvae types and recognition of them. As you know, the *Anopheles* is the major carrier of malaria. The diesel fuel was used to kill the larvae as this would block their breathing.

One might wonder why it was that Bill Schultz, a Navy SeaBee, was trained in malaria control techniques and procedures by the U.S. Army. The reason for that preceded Bill's arrival by almost a year, dating back to the fall of 1943. The story goes something like this.

From March 1942 until September 1943, malaria control operations on New Caledonia and the New Hebrides were under the direct control of the U.S. Navy. However, after a series of skirmishes with local French authorities, specifically on the New Hebrides, malaria control responsibilities were transferred to the Army Service Command and remained so until the end of the war.

Though there had been many conflicts between French authorities and the Navy regarding their malaria control procedures during their first year and a half of occupation, one incident took place that resulted in drastic change of command authority.

Attached directly to the staff of Admiral Halsey was a high ranking Navy malaria control officer by the name of Lt. Cmdr. James Sapero. As Sapero gathered malaria statistics from both the Army and Navy stationed on the New Hebrides, he determined that the highest rates of malarial infection were to be found in those military units stationed closest to French plantations scattered throughout the island. Sapero theorized that large numbers of natives lived on or near the plantations and that the majority of them were more than likely infected with malaria.

As mentioned earlier, the *Anopheles* mosquito is the culprit in spreading malaria from one human to another. In this case, biting the infected natives around the plantations, flying a short distance to nearby U.S. military encampments and passing the disease on to the heretofore uninfected troops. He also reasoned that the highest incidence of mosquito activity occurred between dusk and dawn.

Navy malaria control experts, led by Sapero, further deduced that they could break the cycle by physically moving all of the natives from the plantations located near U.S. military installations before dusk and returning them only after dawn the next day. If the *Anopheles* was most active at night, they would find no malarial natives to bite, hence, the spread of malaria would be thwarted.

The ritual of moving the natives took place every day and would be enforced under the severest penalty. Once moved, the natives were to sleep in shacks constructed by the Army Service Command with transportation duties performed by Army personnel using Army trucks. Obviously, French plantation owners and Army authorities were never consulted, nor was their opinion considered.

When they learned of the Navy orders, the French were, quite naturally, outraged. Their lives were about to be disrupted and disorganization would be the rule of the day. Plantations would be stripped of servants and field hands needed to perform any daily chores after dark. The Army, too, was aghast at the impending strain that would be placed on their Service and Transport commands.

After Admiral Halsey refused to become embroiled in the mess between the Army and the French on one side and the Navy on the other, the French High Commissioner for Espiritu Santo, Monsieur Foucade, lodged a formal complaint with the War Department in Washington, D.C. Response to his complaint was rapid. The War Department issued Circular #178 on September 10, 1943, ordering the Navy to rescind their plan to move the natives. It further ordered that all malaria control operations on the islands would henceforth fall under the direct command of the U.S. Army.

William K. Schultz, CM3/c, USN Malaria Control CMB538, SeaBees. My day was spent about the same as the others you mentioned in MCU Cactus, that is, loading my tank with diesel oil and setting off on foot looking for anything that had water standing in it, such as rotten coconut shells, tree crotches, rotted, fallen trees, and the like. It would take me about a week to ten days to cover my area and I am sure there must have been a few spots I missed. As you can imagine, it is quite difficult to try to walk in a straight line through jungle conditions, then try to figure out when you had walked a mile, or if you had even gone in a straight line. Sometimes your oil would run out and you had to walk back, refill the tank, and start all over, but usually on a different route than the one you originally took because it was difficult to find the place you had last left off. Another thing, on most days, I would refill that tank three or four times, and that made it even more difficult to keep track of where you were.

Over a period of time, one did get to recognize landmarks such as certain trees; some of which were huge. There were lime trees with good fruit, orange trees with huge oranges but not good tasting—they were very sour. There were lots of banana trees with large bunches of small fruit, most about four inches long and very tasty. Also, there were a lot of papayas.

My work was very hot and dirty as one would expect, however, I also got to like it since I was my own boss and had no one to report to and was never given any extra duty. I did have a lot of laundry to do because of the oil and the dirt it attracted. Also, I would shower twice a day because the diesel fuel tends to burn your skin.

We did very little drainage work there. It would rain almost every day, sometimes very hard for an hour or so, but in two hours time, you could not tell it had rained except where there was a container sitting around that would hold the water. The soil there was very porous. It was this constant rain that made malaria control very difficult since it washed away the film of diesel oil we had sprayed earlier.

I think at one time we had some aerosol insecticide sprays, but they were not like the spray cans we now have. If memory serves, they had a valve we opened and then vacated our Quonset hut for a couple of hours. We had the old pump spray cans mostly for indoor use, otherwise.

I took as a suppressant for malaria and I did not ever get it while I was there. As I remember, the sometimes made us look a little yellow. To the best of my knowledge, no one in my unit ever did. In fact, we kept the mosquito population in our area to a very low level and we were seldom ever bitten. We slept under mosquito netting and I even had a netting to cover my face, which I sometimes used when I was in the jungle.

As with so many other men assigned to malaria control initiatives on hundreds of South Pacific Islands, William Schultz was stuck with his contribution to the war effort for almost nineteen months before finally returning home. Unlike others in perhaps more civilized areas of operation, however, William needed to be mindful of much more than a mere mosquito bite.

William K. Schultz, CM3/c USN Malaria Control CBMU 538, SeaBees. Half the island we wouldn't even go on. We had our half and there were headhunters supposedly on the other half, so we just stayed close to where our bases were. There was a road there that covered about half of the island and we were told to stay off any of the trails that led from the road itself.

I was never instructed about my conduct if and when I would meet up with any of the natives. Some of them could converse in pidgin English but it was at times difficult to communicate since, as an example, they had no comprehension of time. For example, I had asked one of the natives to make me a shark's tooth sword and tried to find out from him when I could meet him again to collect the sword. He did not know days of the week or months or years. His only reply was, "By-n-by." I also had a cane made and again it was delivered, "By-n-by."

Most of the natives I encountered were out hunting for wild pigs of which there were quite a few. The boars mostly had very large tusks. Sometimes, when I was out spraying, I would hear shouting, then a rush of pigs followed by dogs and five or six natives running by. They had bows and arrows tipped with four to five heavy pieces of wire, and some of them also had clubs or spears. Usually the dogs would corner a pig against a tree, then the natives would kill it the best they could, sometimes even hitting the pig with large rocks. They would then tie the legs together with small vines, put a pole between the legs and carry it off on their shoulders.

A few times I came across what must have been a python, since I would guess they were around twenty feet in length. I would see them in the trees; however, I was never bothered by these. I did always carry a Zippo lighter with me and with this could turn my sprayer into a small flamethrower in case I was in danger. I never had to do this, but I did practice at it.

Chapter 11

Homeward Bound

May 1944-September 1945

Robert Michel PhM3/c, USN Malaria Control Unit Cactus. Towards the middle of May, 1944, Dave and I were finally ready to go home and our orders were processed, transferring us from Pacific duty back to the States. Once again Dave and I were back together as we arrived in Noumea, New Caledonia, to be processed out for the trip home.

Dave Minton PhM3/c, USN Malaria Control Unit Cactus. While we were stationed there at Noumea, they had a bunch of barracks where they had Bob and I stay. The buildings had a big parking lot in front of them and I can still remember the crushed rock they used to pave it.

The only incident of any note that happened while I was there was that I got into a fight. I was really drunk and got into it with another guy. He was pretty tight too. I remember it was a real brawl; we were punching each other and had knocked each other to the ground and were really trying to kill each other. We probably would have killed each other except that we were arrested and thrown in the brig. We were eventually let go because we were both too drunk to even know what we were doing.

Robert Michel PhM3/c, USN Malaria Control Unit Cactus. Almost two years earlier, when Dave and I had volunteered to head up into the islands where the action was, and left Noumea for Efate, we had been ordered to leave our sea bags with all of our Navy uniforms back there in storage. Believe it or not, our sea bags and uniforms were still there when we were ready to go home.

We were soon to find out that there was yet another problem we had to face. At that time in the war, there was no available transportation back to the States for anyone other than the sick and the wounded. Dave and I had to go to work again to solve that problem.

Bill McLaughlin, PFC, 221st Field Artillery Battalion, Americal Division. We had spent nearly eight months on Fiji, and then we were shipped off to Bougainville in early June 1944, the largest of the Solomon Islands.

But to answer your questions, I was sent to school by the Army to study many things over time, among them, malaria control. I, in turn, passed on the info I had learned to the men in my unit. This took place on Bougainville, another hot bed of malaria. The malaria control duties I had to perform were in addition to my regular duties.

On Bougainville, our sector was the bivouac area in the camp where we had dug in our howitzers. A small stream ran behind the camp in a gully, and this we controlled for malaria—cutting brush along the edges. No special equipment was given us for this job. We just used machetes to cut the brush. We were given by the medic and most of us ignored it. When someone would come down with an attack of malaria, though, he would eat a fistful of those pills so it would show up in his blood when he was tested.

We had mosquito nets on our beds, but we had no lotion to apply to our skin for mosquitoes until the last months of the war, almost a year later.

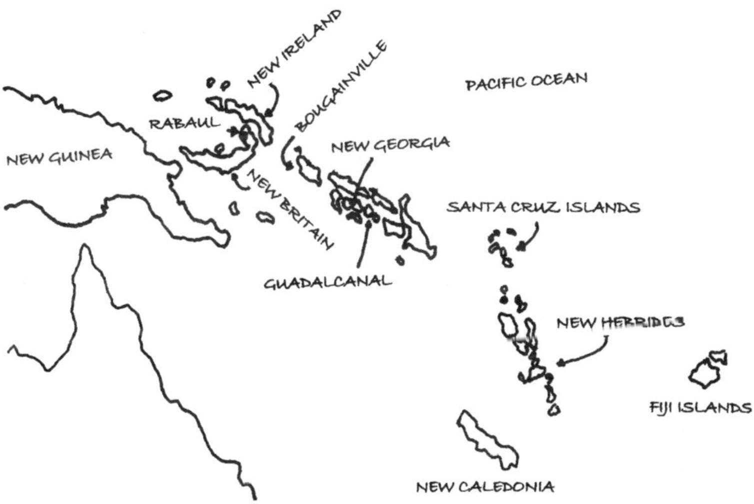

BOUGAINVILLE

Robert Michel, PhM3/c USN Malaria Control Unit Cactus. We learned that the quickest way home for us was to be signed on a ship as ship's company, crewmen. With a little effort, we were able to talk our way onto a ship heading in the right direction. As a PhM2/c, I ended up being in charge of the ship's psycho ward.

The psycho ward that I was assigned to also had a cell with bars where violent patients were kept. The sad thing about it was the one guy who was kept in there was suffering from what was called at the time, battle fatigue. As I talk about this now, I suddenly remember other things I thought I had long ago forgotten.

We had just set sail from Noumea, New Caledonia. One of our patients somehow had gotten loose and jumped over the side of the ship. He said he wanted to go back to Noumea and that's what he intended to do when he went over the side. The ship never even slowed down for him for fear of being attacked by a Jap sub. We had no destroyer escort. As I remember, we were much too far from shore for him to have made it safely back. But who knows? Maybe he did.

The one guy who was being held in the cell of the psycho ward was violent. He had nothing on but a pair of skivvies. He just sat on a mattress on the steel deck with his boondockers sitting beside him. When I took his food to him, I had a rolled up, wet dishtowel doubled over with one end tucked under my belt. I had been ordered to just beat the shit out of him with that towel if he ever became violent and had to be subdued. I never did have to use that towel but I am sure I would have if I had needed to. What a way to treat a mental illness.

Fortunately, most of the men, and I don't remember how many of them there were in that ward, were nonviolent and didn't cause any trouble at all. They were just a little bit crazy. They were living in a different world. One can only imagine what they had gone through to wind up in there in that condition.

That, however, is how Dave and I got back to the States that docked in San Francisco in May 1944.

Dave Minton PhM3/c, USN Malaria Control Unit Cactus. I don't remember the name of the ship we came back on nor a whole lot of detail about the trip except for a couple of things. There was a whole ward on the ship filled with Japanese prisoners who were all in pretty bad shape, wounded, a lot of gut wounds and the like. They were really bad off. I remember I worked in the ward and worked on a lot of those Japs.

The trip back to the States, though, was a lot better than the one we had going over. On the way back, we didn't have to worry about Jap subs and shit like that; no worries about Jap planes either.

Bob and I came home on what they called the "point system." Bob and I had enough points accumulated and that's how we got on that ship. A certain number of points were assigned for where you had been, what you had done, and other stuff like wounds, medals, that type of thing. I just don't remember all the details now.

The Golden Gate, San Francisco , and Home

Robert Michel PhM3/c, USN Malaria Control Unit Cactus. A thrilling thing happened when we were approaching San Francisco. As our ship was approaching the Golden Gate Bridge, we were all out on deck to finally see the good old USA again. We were such a rag tag bunch of guys. Most of us were in old, dirty clothes that were almost rags. Up on the bridge, people stopped their cars and looked over the side and waved at us. I guess we were one of the first ships to return from the South Pacific. It was one of the most exciting moments in my life. What a different greeting we received compared to that of the men who came back from Nam. Regardless of anything, our country really shit on those guys.

When we finally pulled in to San Francisco, Dave and I had to attend some sort of muster at the base on a parade ground. We were standing around waiting to get our liberty cards or some such thing.

At the end of May 1944, San Francisco was a city that had been prepared for war for almost three years now. Flooded by military personnel heading off to the Pacific from her crowded airfields and bustling docks, San Francisco was to slowly change from the busiest port of departure to the busiest port of entry as victory in the Pacific was reached. Over the next two and a half years, almost a million servicemen would be following Bob and Dave home; their first sight, the Golden Gate Bridge. Bob and Dave, however, were among the first to return with the war's most gruesome battles yet to be engaged: Saipan, Iwo Jima, the Philippines, and Okinawa.

The San Francisco that Bob and Dave came back to in '44, however, was a city that had lived under the threat of Japanese air attack, whether real or imagined, for almost two years, since December of 1941. Blackout regulations were yet rigidly enforced all up and down the West Coast and wouldn't be relaxed for another 5 months. The Pacific Coast Warning Service of the Fourth Air Force monitored air activity in an underground air raid warning center in the basement of City Hall. Citizens of Japanese descent living in the area prior to the war would not be allowed to return to their homes for almost two more years. Ration tokens and stamps were a requirement when purchasing any foodstuffs, clothing, or gasoline. Prices were set and monitored by the OPA on literally every item a person may wish to buy or sell. The average person could not buy an automobile like you can today, there just weren't any for sale. If you did have a car, parts and particularly tires were nonexistent.

Wartime San Francisco may sound like a pretty grim place for those of us unfamiliar with life during the war years, but to Bob and Dave, being back in the States was all that mattered. Having survived the past two years in the South Pacific, they were ready to celebrate. Close the windows and lock the doors, no shortages or blackouts were going to inconvenience Bob and Dave. In fact, they didn't even notice. Considering where they had just come from, Bob and Dave knew they were as close to heaven as the real thing must be. There were real American women and they were everywhere they looked. There was genuine Scotch; a little tough to get, maybe a little expensive, but definitely not out of reach. There were restaurants with seafood, steaks, and even hamburgers. And, with unlimited liberty while in town, Bob and Dave knew just how to parlay their status as heroes of the South Pacific and the glorious 1st Marine Division to their best advantage.

Dave Minton PhM3/c, USN Malaria Control Unit Cactus. When we landed in San Francisco, we were stationed out on Treasure Island. It didn't take Bob and I long, though, to hook up with a couple of girls and shack up with 'em.

Just nine years earlier, in 1935, Treasure Island, California, existed only in the minds of forward thinking San Francisco city planners. They had decided collectively that they would hold a World's Fair in 1939, which they dubbed the Golden Gate International Exposition. The Fair would be the perfect venue to showcase the engineering masterpieces so recently completed in their fair city, the magnificent Golden Gate Bridge and the San Francisco and Oakland Bay Bridges. They realized that a fitting site would be needed to adequately stage the event and properly impress a couple of hundred thousand world visitors. City fathers agreed, a man made island constructed on the shallow shores of Yerba Buena island, smack in the middle of San Francisco Bay between the city of San Francisco and Oakland, would provide the perfect location. As an afterthought, the island would provide an excellent location for a new airport serving communities on both sides of the Bay after the World's Fair departed, and what better place to construct an airport than right at the halfway point of the San Francisco and Oakland Bay Bridge already connecting communities on both sides of the Bay.

Calling upon the Federally funded WPA for manpower, construction was started immediately. Dredging mud from the bottom of the Bay and dragging fill from the Sacramento River delta just to the north, an island emerged two years later from the inky waters of San Francisco Bay. It covered 403 acres of land and was made up of over 29 million cubic yards of imported sand and gravel, along with over 259,000 tons of rock that were used to wall the island, thereby preventing damaging erosive effects of the Bay. The newly constructed dot on the map was quickly named Treasure Island, a not so subtle reference to the gold filled sand and dirt carried down from the Sierra Nevada Mountains by the Sacramento River used in its construction.

The Pageant of the Pacific, as the theme of the World's Fair was nicknamed, saw over 200,000 visitors walk the exhibits on the newly opened Treasure Island over the next two years and hundreds of departing and arriving flights of the Trans-Pacific Pan American China Clippers carrying mail and passengers to and from destinations in the Orient.

By February 1941, and war with Japan more a matter of when rather than if, the United States Government leased Treasure Island from the City of San Francisco. Just two months later, it was seized and occupied by the United States Navy, bringing with it a change in name to Naval Station Treasure Island that would last until Congress closed the base in 1993. Approval of the transfer was finally agreed upon by officials of San Francisco when the Navy offered in trade a parcel of land in South San Francisco which ultimately became the site of the present day San Francisco International Airport.

By the time Bob and Dave arrived at Treasure Island to be processed back into an East Coast Naval medical unit, Treasure Island had long been the home to the headquarters for the 12th Naval District and had served the past three and a half years as the Gateway to the Pacific. Hundreds of thousands of troops and hundreds of loaded ships and planes bound for Pacific battlefields had shipped out from her docks. Now that the war was beginning to wind down, Bob and Dave became some of the first returning troops that would swell again to hundreds of thousands as Treasure Island became the gate back into the States for returning servicemen.

Treasure Island's role as the Gateway to the Pacific would continue for another three and a half decades after World War II, peaking again during the 1960s and early 1970s. As the war in Vietnam escalated, thousands of U.S. Marines and sailors once again passed over her shores on their way to and from the Pacific Fleet and the front lines of Vietnam. The author was among those thousands, who in late 1969, was ordered to the Seventh Fleet by way of Treasure Island. Boring though it may have been to be a transient in the modern military awaiting assignment to whatever duty station may be ordered, many nights were spent that fall watching the fires started on Alcatraz by the American Indian Movement or partaking of the night life of San Francisco on a welcome liberty.

Treasure Island was only a short layover for Bob and Dave as it would be for so many to follow. After two and a half years in the Pacific, Bob and Dave would need new uniforms, probably a physical and some updated inoculations, and the paperwork required to see them on their way to new duty stations. In the meantime, as celebrities of a sort just back from the Pacific, Bob and Dave were granted almost unlimited liberty, temporarily stationed at Treasure Island only long enough for the Navy to catch up on their paperwork. Two or three weeks, and Bob and Dave would be on their way to their next duty station, this time, hopefully, closer to their respective hometowns.

Robert Michel PhM3/c, USN Malaria Control Unit Cactus. We had our sea bags and equipment with us and left them on the ground where we were standing when we went to the front to get the liberty card. I also had a small, blue canvas carry bag in which I had 35mm film negatives stashed along with a pearl I had found on a dive in the ocean off the Floridas. I also kept a lot of personal stuff in there and a carton of cigarettes. I had emptied out a tin of tooth powder and used the empty can to hide the film and the pearl.

I left the bag for only a moment or two, taking my turn getting the liberty card. When I got back, it was gone. Some son of a bitch stole it. The guy probably threw everything away except the cigarettes. That pearl was a real beauty. Probably not worth anything, but to me it was priceless. It resembled a teardrop in shape, about ½ to ¾ of an inch in length. I had planned on making it into a pendant for my mother. Ah well.

Later that day, we picked up new uniforms and headed out on the town. I insisted that our first stop be an oyster bar for some raw oysters and beer. You see, every year during the '30s on my birthday, my mother would treat me to an afternoon at Radio City Music Hall in New York City. Before we went to the show, we would always stop at a little Chinese restaurant located on the second floor of a building on 6th Avenue right near Radio City. My mother loved oysters on the half shell and they were fairly inexpensive at that time. I have some wonderful memories of watching her eat those every year on my birthday. She probably used up what little money she was able to save during the year for this one special day. I would watch my mother eat and savor every one of those oysters and would ask her for one to try myself. I was never able to swallow even one.

One day on Guadalcanal, out of the blue, like a pregnant woman, I had the most unbelievable craving for oysters on the half shell. That craving stayed with me until I got back to the States. My first stop had to be for oysters on the half shell. I've loved them ever since.

After getting a little tight and getting a hotel room, we went back out on the town and visited one bar after the other and finally staggered into a liquor store to buy a bottle. The owner wouldn't sell us one because he said we already had too much. By the way, there was a third guy with us named Ely Riley. He was a big Irishman and we took him along for protection. Anyway, after we came out of the store, Dave said, and I will never forget his exact words, "Look at my little baby." Under his jumper, he pulled out a quart bottle of booze he had stolen off the rack. It gets better.

After the three of us downed most of the bottle, we passed in front of a burlesque house and Dave insisted we go in. So he bypasses the ticket booth and goes directly to the ticket taker and points to me like I have the tickets. I went in behind Dave and pointed to Riley like he had the tickets. Riley, who was one big guy, grabbed the ticket taker by the front of his uniform and said, "We gave you our tickets."

So now here we are inside the burlesque house and instead of going down the center aisle, Dave goes down the side aisle all the way to the front. There is a pair of steps leading up to a box on the stage with a railing around it. Of course, Dave goes up to the box and we follow.

So there we were, three drunken sailors practically sitting on the stage. When the star came out and saw us, she naturally thought we had paid extra bucks for our seats, so she started to give Dave, who was in front, his own special performance. She would bump and grind almost in arms reach and he would make a lunge for her and miss. The audience loved it, so she kept it up.

It was a blast, though, until the SPs showed up and hauled us out of there and into the brig, where a salty Chief Boatswains Mate was in charge. We gave him a long song and dance about our two years with the 1st Marine Division on Guadalcanal in the South Pacific, so he let us go with a warning to go back to the hotel and sleep it off. He even let us keep what was left of the bottle. That was just our first day back in the States.

While in San Francisco, Dave and I eventually met a couple of girls. One of them, Nathiel Austin, invited us to her house and took us around to the back yard where there was this old Ford Model A coupe parked. She told us it had been there for years and wouldn't run. Never let it be said that I couldn't make a machine run. I finally diagnosed the problem as fuel starvation. I found a thing on the block that looked like the fuel pump. I got some tools and took it off the engine block. It was very rusted and all seized up. I forget exactly how I did it, but I got this mechanical pump to work and reinstalled it. Lo and behold, the damn car started. No plates on the car or any registration, but Dave and the girls and I rode all over San Francisco in that old jalopy and never got stopped once by a cop. We had a ball. Shortly after that, it was time for us to head back to New York City and then home.

When Dave and I left San Francisco to go home to the East Coast, it was by train; the old fashioned kind that burned coal to make steam. The soot from the coal permeates everything, and its not too long before you have it all over you. There were no seats available on that train—not one—so we either stood all the way or sat down in the aisle. Those old trains had hammock type overhead places to put your luggage in. At night we would climb up over the seats and get into those hammocks and sleep.

We met a couple of girls on the train who lived in Chicago, and we became quite friendly with them. When we arrived in Chicago, Dave and I decided to get off and have a date with the girls. We had their address, so we were ready for bear.

In the spring of 1944, there was little doubt in the minds of Democrats, or for that matter, most of the nation's populace, that Franklin Delano Roosevelt would run for an unprecedented fourth term as president of the United States. Roosevelt had only recently written a somewhat humbly worded letter to the Democratic National Chairman that he would again run for president reluctantly as a good soldier. Delegates to the Democratic Convention overwhelmingly nominated Roosevelt and tapped Senator Harry S. Truman as his vice presidential running mate, setting the stage for Truman's ascendancy to the presidency on the death of FDR just four months hence.

When Bob and Dave boarded the California Zephyr bound for New York, their thoughts were of home. They had just spent the better part of two years overseas, much of that time with the First Marine Division in hellish conditions on Guadalcanal. Still shy of twenty one years of age and yet unable to cast their vote for president, neither Bob nor Dave knew or cared that much what historical events were taking place that summer in Chicago as they made their way back home.

While Bob and Dave were catching up on the good times they had missed out on, the Democratic Convention had a stranglehold on the Windy City, occupying every hotel room in town and tying up the city's public transportation. The mood in town was jubilant as the war in Asia was going well and U.S. troops had just stormed ashore on the beaches of Normandy, opening a second front in the war against Hitler. The end of the war was truly in sight.

Dave Minton PhM3/c, USN Malaria Control Unit Cactus. When we got to Chicago there was a political convention going on, Republican or Democratic, I don't remember. That town was really crowded, I mean, there wasn't a room to be had anywhere. There was just no place for Bob and I to go.

Robert Michel, PhM3/c USN Malaria Control Unit Cactus. Now picture this. Two swab jockeys in filthy, wrinkled, hot dress blue wool uniforms; no bathing for three days. We were able to shave only once on the whole trip. It's June and hot as hell in Chicago. We got off the train and went to the taxi circle in front of the station only to find a line a mile long of people waiting for a cab.

There was a policeman directing the cabs and the waiting people, so we figured we would never get a cab. Now remember, it's a hot summer day in Chicago. I told Dave, "I'm going to put on my pea coat and start shivering and sweating. You go out to the cop and tell him we just got back from the islands and your buddy has malaria—can he help?"

Dave did, and I did, and the cop whistled a cab for us. He told the cabby to take us to the nearest hospital, but after we got in the cab we told the driver to take us to a hotel rather than a hospital. He just laughed.

One of the presidential conventions was in town and there wasn't a hotel room available for fifty miles. The cop had told the cabby to take us to a hospital and we were trying to get the cabby to take us to a hotel. Finally after much persuasion, the cabby finally did.

I told Dave to pull the same thing with the desk clerk that he did with the cop, so both of us walked into the lobby. I stayed near the door and Dave went up to the front desk. Pretty soon Dave called me over. The clerk had told Dave that there was absolutely nothing available when the hotel manager walked out and got into the conversation. He told us there had been a cancellation for the night and that he would be able to let us use that room for the night.

Dave Minton PhM3/c, USN Malaria Control Unit Cactus. I mean to tell you, Bob looked really bad. You know, here's a guy standing out on the street with malaria for God's sake. I just told Bob to shiver. If he shivered looking that bad, I figured the guy in the hotel behind the desk would feel sorry for us, which he did and we got a room. That was at the LaSalle Hotel. The old LaSalle later burned down, I heard.

Robert Michel PhM3/c, USN Malaria Control Unit Cactus. Again, picture this. Two dirty swabbies in hot wool blues, each carrying a sea bag and hammock; one shivering in a pea coat with the temperature outside around 80-90 degrees.

The manager had a couple of the bell boys take our bags and escort us up to our room. This was not a room…it was a suite of rooms. We couldn't believe it. There was even a liquor cabinet filled with booze.

To make a long story short, we showered, shaved, and had a wonderful time with the two girls we had met on the train. The next morning when we checked out, our bill was a total of $2.00. They wouldn't charge us for the room because they said it was already paid for. The two bucks was for a change of linen. Can you believe that? I think back on those times and how the civilians tried to help the servicemen in any way they could. They went out of their way to show their appreciation of us.

We caught the train that took us back to New York. Dave went his way and I took the ferry over to Wehauken, New Jersey, to catch a train to Bogota, a town near my home. As soon as we had gotten to New York, I had called my mother and told her I would get the train and meet her in Bogota and that I would be home soon. When I got off the ferry and was walking toward the trains, who should I run into but my hometown, serious girlfriend, Martha. Martha Hanway, from Ridgefield Park, was on her way to work in New York. To this day I can't figure out what happened. I tried to convince her to take the day off from work and come with me, but she insisted she had to go to work. I think the fact that she wouldn't take the day off to spend time with me after two years had a lot to do with why our relationship never matured into marriage, although we did see a lot of each other for the next six months or so.

But the story isn't over yet. After I saw Martha onto the ferry, I started back toward the trains when I heard, "Hey, Bob." I looked up, and there was my stepfather in the cab of the train I was going to get on. His name was Harry Fish. Harry was more a father to me than my real father, and I really loved that guy. He was an engineer on the New York Central and had started on the railroad when he was 16 years old. He had me climb up into the cab of that coal burning steam engine and that's the way I rode to Bogota, New Jersey, the day I got home from the South Pacific and started on my first, thirty-day leave since my R & R in New Zealand the year before.

When Bob stepped off the train in Bogota, New Jersey, that morning early in the summer of 1944, he, like many thousands of other servicemen rotating back to the States from battlefields in the South Pacific, Africa, and the Mediterranean, was suffering from an acute case of malaria. We learned early on in Bob's story just what the implications of his illness might mean, not only in terms of his own health, but quite possibly to an innocent and until now uninfected population in his own hometown.

There are eight subspecies of the *Anopheles* mosquito living and breeding in the continental United States, two of which are capable of transmitting the malaria *Plasmodium*. Conditions for a possible resurgence in the spread of malaria in the United States were now in place: the *Anopheles* mosquito, an infected reservoir (Bob and company), temperatures over 60 degrees, the summer months of 1944, and a susceptible population. The threat of indigenous transmission of malaria was once again at an all time high.

In 1942, only four cases of malaria were reported in military personnel in the state of New Jersey, with sixteen cases documented in the civilian population. In 1944, seventy-eight military cases of malaria were tallied with just eight civilian cases recorded. One year later, in 1945, as the war was winding down and the rush was on to bring the boys home, 1,397 military men were reported to be infected with malaria and civilian cases increased to 15 in the state of New Jersey alone. Unbeknownst to Bob, his mere presence now posed a potential health risk to family and friends alike.

For the next 4 years, Mrs. *Anopheles* waged a malarial attack on parts of New Jersey and across the country until state and local Malaria Control and Health Organizations once again reorganized to fight the mosquito vectors in an effort to minimize the growing number of cases of malaria. By 1950, statistics showed a steady decline in imported cases of malaria like Bob's and a termination of any new natural transmission of the parasite. By

1951, with stringent malaria control, eradication, and testing procedures firmly in place, malaria had once again been completely eradicated from the state of New Jersey. Similar success stories were pouring in from all over the country. For over 20 years, New Jersey and many other states across the country remained totally malaria free.

Robert Michel, PhM3/c USN Malaria Control Unit Cactus. After my thirty-day leave was over, I was transferred to a ninety-day wonder school in Asbury Park, New Jersey. It had been a hotel before the war and had been taken over by the Navy for the purpose of training recruits to be officers in ninety days.

As a PhM1/c, I was the ranking medical petty officer on board and had a few corpsman under me who worked the sick bay where we had a few beds. My immediate superior was a nurse who had the rank of full Lieutenant. For some unknown reason, she and I had a terrible personality conflict. She would do anything possible to get me to do something that would bring me before the mast—to get court-martialed. After being there for a few months, I realized I had to do something to get away from her or I would end up in the brig.

The straw that broke the camel's back was when she ordered me to sharpen hypodermic needles. I told her I would have one of the corpsman take care of it. She said she wanted me to do it and I refused. By this time I considered myself a sea lawyer who knew a little bit about regulations, so I told her that a PhM1/c was not required to do things like that and that I would have it done as she requested. I also now knew I was in deep trouble.

It was a Friday and I had liberty that weekend. I could feel an attack of malaria coming on, so I talked a shipmate of mine into taking me home to Hackensack. I had a 1937 Packard Sedan at the base. I cleaned out my locker of everything I had and put all my stuff in my car and when liberty started, my buddy and I got into the car and headed for Hackensack. I knew that with the malaria coming on I had a way of getting out of that nurse's clutches.

When we arrived at my home my buddy took a bus to wherever he was headed and I walked into the house. My mother saw me for the first time with malaria. She was in a state of panic, seeing me shivering and sweating the way I was. The symptoms now were full blown. When I finally assured her it wasn't as bad as it looked, I told her to call the 3rd Naval District office in New York and tell them her son had malaria and must get to a hospital. With that I got into bed and waited for that white Cadillac Navy ambulance to arrive, which I knew would happen. I also knew that they would take me to St. Albans Hospital on Long Island where I had a lot of friends, including doctors. Once there, I knew I would be able to get one of the doctors to have me transferred there, to Ship's Company. You see, I was a sea lawyer!

To make this long story short, everything went as I assumed it would. After getting out of the hospital ward I was transferred to St. Albans Naval Hospital. I never saw or heard from that nurse again.

Show Biz and the U.S. Naval Hospital St. Albans, New York 1944

Robert Michel, PhM3/c USN Malaria Control Unit Cactus. While at St. Albans I met a guy by the name of Ed Bykowski, also a PhM1/c who had been injured on the Vincennes when it was sunk off Guadalcanal. He and I organized a variety show for the hospital patients. We had a twelve piece band, a female vocalist, Ed, who was a comedian, and yours truly as the announcer; all from the staff at St. Albans. We formatted the show like the popular radio variety shows of those days. We would put on the show once a week in the hospital auditorium.

Somehow, Mutual Broadcasting got wind of the show and came out to see it. They liked what they saw and decided to put us on network radio once a week to sell war bonds. Mutual Broadcasting arranged for celebrity guests and we were able to keep that show going for a few months or so.

One of the celebrity guests happened to be the famous actress Talullah Bankhead. We became pretty good friends and she would leave tickets for me at the theater on Broadway where she was appearing in a show called “Foolish Notion.” If I were in New York on liberty and had nothing to do, I would sometimes go to the theater and watch her show. After some time, Talullah asked me if I wanted to be an actor and I said yes, so she invited me to go on the road with her and the play to learn acting from the ground up. She said I wouldn’t get paid, but that the show would provide food and lodging and she would teach me the theatah, Dahling.

What an opportunity, I thought! But, in talking it over with my mom, she said she didn’t think the wicked stage was where she would like to see me. So, after thinking it over for a few days I decided to do what she wanted me to do and that was to decline the invite. Who knows what might have ever come of it? Now, this might sound like I was a mama’s boy. I wasn’t. I just loved my mother very much and didn’t want to hurt her in any way after all the sacrifices she had made for my brother and I.

As a result of that show on Mutual Broadcasting, I learned about an audition that radio station WNEW was holding for an announcer. It was a new show they were putting together featuring the musicians from the Fred Waring Pennsylvanians, who were in the service. They had gotten all of the guys together and were looking for a serviceman announcer.

I went to the audition and I got the job. Once a week I would go up to a studio on 67th or 69th Street to do the show. The show lasted a few months and somehow, as a result of that show, I met an announcer from NBC whose name was Jack Costello. Jack later got me a job with NBC as a page.

I guess now is as good a time as any to tell you how I got out of the Navy before the war with Japan was over.

As I mentioned earlier, I considered myself a pretty good sea lawyer, so one Monday morning when I got back from liberty and was a little hung over and totally fed up with the Navy and all its chicken-shit regulations, I decided to act. I was walking down the main hall of the hospital when I passed the office of the captain who was head of psychiatry. I figured what the hell, and walked into his office. He was sitting at his desk and looked up at me and asked, "Can I help you son?" I told him yes, he could get me out of this f@#$! latch-up. He didn't bat an eye. He just took my name and rank and said he would look into it.

By the end of the week I was on my way OUT.

By now it was June 1945, and I was finally being discharged from the Navy at Lido Beach, Long Island. At the time, I was a Pharmacist Mate First Class. I was standing in a line waiting to be processed out and who should be sitting at the desk doing the processing work but DAVE! Can you believe all these coincidences with Dave? When I saw Dave sitting at the desk, he looked at my papers, stamped and signed them, and I was at last discharged from the Navy with a Battleship Discharge.

V-J Day

On Sunday, September 2, 1945, almost four years after Bob Michel first heard the news of a Japanese sneak attack at Pearl Harbor on a patient's radio at Quantico Marine Base, Japanese officials signed formal surrender documents aboard the battleship USS Missouri anchored in Tokyo Bay. General Douglas MacArthur signed on behalf of all of the Allied Powers and Fleet Admiral Chester Nimitz placed his signature on the documents as the representative of the United States. Just that morning, U.S. Army forces landed at Yokohama, Japan, to begin the occupation and rehabilitation of a defeated empire. Celebrations worldwide had been taking place almost nonstop since the first formal confirmation of Japanese capitulation had been received two weeks earlier on August 14th. Fours years of total war had finally come to an end, albeit with the destruction of Hiroshima and Nagasaki by atomic bombs and the threat of more to follow.

Excerpt from Department of State Bulletin 13
September 2, 1945
Address of President Harry S. Truman

The thoughts and hopes of all America, indeed of all the civilized world, are centered tonight on the battleship Missouri. There on that small piece of American soil anchored in Tokyo Harbor, the Japanese have officially laid down their arms. They have signed terms of unconditional surrender.

Four years ago, the thoughts and fears of the whole civilized world were centered on another piece of American soil, Pearl Harbor. The mighty threat to civilization which began there is now laid to rest. It has been a long road to Tokyo and a bloody one.

We shall not forget Pearl Harbor.

The Japanese militarists will not forget the U.S.S. Missouri.

The evil done by the Japanese warlords can never be repaired or forgotten. But their power to destroy and kill has been taken away from them. Their Armies and what is left of their Navy are now impotent.

To all of us there comes first a sense of gratitude to Almighty God who sustained us and our Allies in the dark days of grave danger. Who made us to grow from weakness into the strongest fighting force in history, and Who now has seen us overcome the forces of tyranny that sought to destroy this civilization.

God grant that in our pride of the hour we may not forget the hard tasks that are still before us; that we may approach these with the same courage, zeal, and patience with which we faced the trials and problems of the past four years.

Our first thoughts, of course, thoughts of gratefulness and deep obligation, go out to those of our loved ones who have been killed or maimed in this terrible war. On land and sea and in the air, American men and women have given their lives so that this day of ultimate victory might come and assure the survival of the civilized world. No victory can make good their loss.

We think of those whom death in this war has hurt, taking from them husbands, sons, brothers, and sisters whom they loved. No victory can bring back the faces they long to see.

And so, on V-J Day we take renewed faith and pride in our own way of life. We have had our day of rejoicing over this victory. We have had our day of prayer and devotion. Now let us set aside V-J Day as one of renewed consecration to the principles which have made us the strongest nation on earth and which, in this war, we have striven so mightily to preserve.

Those principles provide the faith, the hope, and the opportunity which can help men improve themselves and their lot. Liberty does not make all men perfect nor all society secure. But it has provided more solid progress and happiness and decency for more people than any other philosophy of government in history. And this day has shown again that it provides the greatest strength and the greatest power which man has ever reached.

We know that under it we can meet the hard problems of peace which have come upon us. A free people with free Allies, who can develop an atomic bomb, can use the same skill and energy and determination to overcome all the difficulties ahead.

Victory has its burdens and its responsibilities as well as its rejoicing.

But we face the future and all its dangers with great confidence and hope. America can build for itself a future of employment and security. Together with the United Nations it can build a world of peace founded on justice and fair dealing and tolerance.

As president of the United States, I proclaim Sunday, September 2, 1945, to be V-J Day, the day of the formal surrender of Japan. It is not yet the day for the formal proclamation of the end of the war or of the cessation of hostilities. But it is a day which we Americans shall always remember as a day of retribution as we remember that other day, the day of infamy.

From this day we move forward. We move toward a new era of security at home. With the other United Nations we move toward a new and better world of peace and international goodwill and cooperation.

God's help has brought us to this day of victory. With His help we will attain that peace and prosperity for ourselves and all the world in the years ahead.

Robert Michel, PhM1/c USN Malaria Control Unit Cactus. When I was discharged, I spent a few weeks hanging around the house and trying not to say the word f@#! too many times in one sentence. You, know, trying to get used to civilian life again. I soon realized I had to get a job and started working for the Bendix Company, located near Teterboro airport a couple miles from my house. I worked there as a guard and walked around the place punching a clock. One day, while at work, the news came over the radio that the Japs had surrendered and that the war in the Pacific was over. That was the end of that job. I ran to my locker, changed clothes, turned in the .38 and headed for home. When I got there I changed clothes again and put on my uniform, got on a bus to New York City and 30 minutes later I was with thousands of other servicemen in Times Square. What a celebration!! Every girl you met wanted to kiss you. Every civilian wanted to buy you a drink. And we let them! It didn't take very long before I was in the arms of a girl and on the way to somewhere. That night is a little fuzzy in my mind except I do remember waking up the next morning with a bad hangover and wondering where in the hell I was. I still don't know. I ended up on the subway 'A' train, then the bus at the Washington Bridge, and then home. For me, the war was now finally over.

When President Truman delivered his address to Congress that night, and Bob was celebrating victory in Times Square, there were still millions of American servicemen spread all over the globe that were just itching to get home. Many hundreds of thousands were in the Pacific and had been poised for the invasion of Japan when the atomic bombs were dropped, ending that horrible possibility. Many, not having accumulated either the required time in service or the number of points needed to rotate home, were assigned various duties in their respective theaters of operation.

These duties ranged from overseeing the surrender of enemy combatants, many hundreds of thousands of whom were still armed and unaware of the surrender, to handling the logistics at hundreds of military bases, to dealing with the rush of supplies still flowing through the pipeline from the States.

Marine Sergeant, Section Leader 75mm Pack Howitzers, I/3/11. When the war ended in August, I was on Guam and our next landing had been scheduled for Japan itself. When they dropped those two bombs, it was all over for those of us in the line of fire. You hear a lot of criticism today about us using the bomb on Japan but, never do you hear that from those of us who would have had to invade Japan. The closer we got to the home islands, the worse it got.

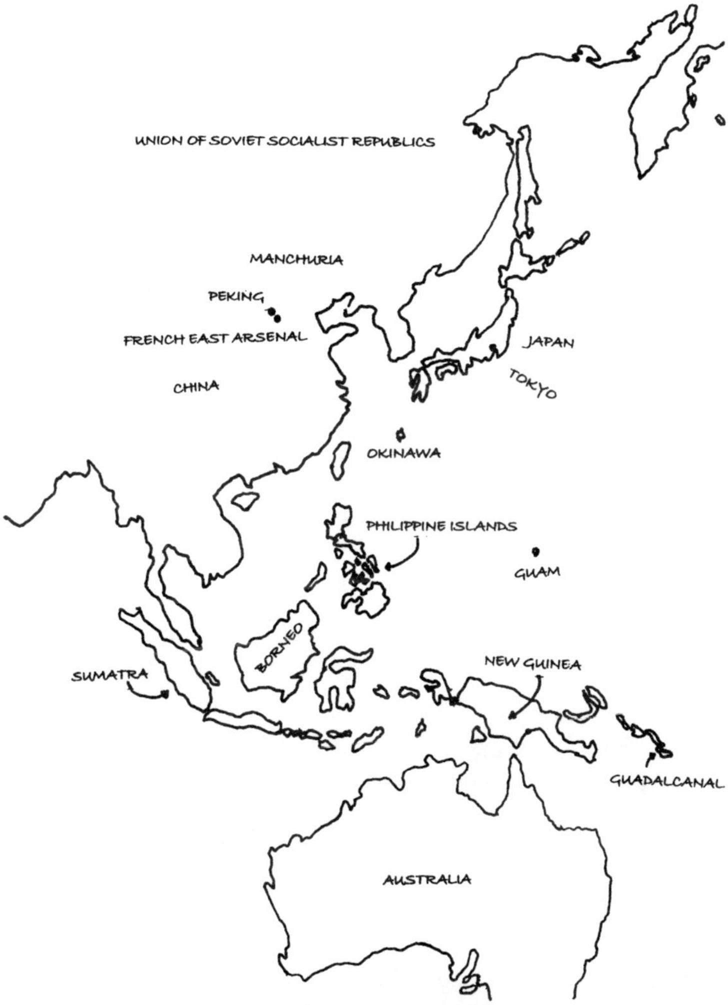

CHINA 1945

Unknown to our Marine Sergeant, and probably most of the fighting men still in the Pacific at the time of Japan's surrender, was the fact that the United States had sustained over 292,000 men killed in action thus far in the war. One authority estimated that an additional 92,500 U.S. soldiers would have been killed if we had invaded Japan's southernmost island of Kyushu alone, scheduled to kick off November 1, 1945. It comes as no surprise then, that when asked how they felt about Truman giving the go ahead to drop the atomic bomb on Hiroshima and Nagasaki, most surviving servicemen who were in the Pacific at the time will tell you it probably saved their lives.

Marine Sergeant, Section Leader 75mm Pack Howitzer, I/3/11. In November 1945, I was sent to China as a replacement back into the 1st Marine Division because I still had almost a year left to go on my enlistment. The 1st Marine Division was stationed all over China to disarm and process Japanese troops that were still there and to act as a buffer between the Nationalist Chinese and the Communist Chinese. I was sent to the 1st Battalion, A Battery of the 11th Regiment of the 1st Marine Division stationed at East French Arsenal about fifteen miles north of Tientsin. It was like a big fort that was well established, built by the French during the colonial powers days. It even had a cemetery where all the French people had been buried years before.

Our job there was to accept the surrender of almost two full Japanese divisions that were still in the area. Because there were so many guys going home, at that time, there was a shortage of officers there. Many jobs that would normally be an officer's responsibility fell to noncoms like myself. So, I was in charge of bringing in and processing the Japs in the area and sending them back to the coast by truck or train so they could be shipped home. We would disarm the Jap soldiers and take all their weapons and equipment. Everything we took from the Japs we gave to Chiang Kai-Shek's boys; you know, the Nationalist Chinese troops. When any Japs surrendered to the Russians, you know what they did with that stuff? The Russians gave all the weapons and equipment they took from the Japanese to the Reds, Mao Tse Tung's Communist troops.

At the compound there was a Japanese captain, along with two civilian Japanese in their mid to late forties who acted as our interpreters. The interpreters had been educated at Oxford, in England, and were very well educated and spoke perfect English. They told us they were in the import/export business. I asked them if that meant they imported Jap soldiers and exported anything that was any good back to Japan. The Jap captain and a lot of the other troops in the area had been in China so long that they had their families with them; you know, their wives and kids. They had homes there and lived right by the compound we were in.

They were absolutely perplexed whenever they saw any Marines bartering or haggling over an item they wanted to buy from the local Chinese. They used to say, "Why are you doing that? You are the conquerors. You should just take what you want from these people." That's the way they treated them, and they couldn't understand why we didn't do things that way. They would just lop off a few heads and take whatever they wanted. That's the way they treated our guys when they took them prisoner.

No Japanese soldier would accept an order from an enlisted man like myself, nor from a civilian, only from a Jap officer. So I would issue my orders to the civilian interpreters who would then pass the order on to the Jap captain who would then give the order to the Japanese soldiers. When they got the order from the Jap captain, they would bow and squat down and do what they were told. They would stack their weapons, line up for a work detail or whatever else we wanted them to do.

Well here's the story I wanted to tell you—absolutely true. I had been there in China for about three or four months when one day the Jap captain and one of the interpreters wanted to invite me to the captain's home for dinner. I thought I would go, but as far as I was concerned, they were still Japs and, therefore, the enemy, so I didn't really trust them. For all I knew they would get me over there and they would jump me or something, so I decided to take along some protection. I had this corporal who was about

6'4"—a big strapping, tough Marine who was a Ute Indian—go along with me just in case there was trouble.

Well, we went over to this Jap's house and when we walked in we had to take our shoes off. We sat on the floor and there was a little low table that they served everything on. They had those little hibachis, like you see today that they flooded our country with after the war. They served warm sake in little cups; not a glass like we would use, but little bitty cups. The dinner was sukiyaki, I think they called it. It was beef with a very sweet sauce on top. The women were not allowed to sit at the table with us, they just stood back against the wall and served us whenever it looked like we needed something.

Everything was going along fine, when I noticed the Jap captain and the interpreter in conversation between themselves. I had no idea what they were talking about as they carried on. The captain said something to the interpreter and he turned to me and said, "The captain would like to ask you a question, with your permission." I told him to go ahead and ask. The interpreter looked at me and said, "The captain would like to know why you think Japan lost the war."

Well, I was just twenty four years old and had been in the Marine Corps for over six years by that time. I had been through two major campaigns in the Pacific and thought that was an easy question. I don't know how much sake I had up to this point, but I just leaned back, puffed up my chest and said, "That's easy. You guys lost the war because you underestimated the United Sates Marine Corps!" With that the Jap captain and the interpreter jumped away from the table and got on their knees in front of me bowing and scraping. "You right, Marines the greatest warriors of all." I felt a little taken back by all this and told them to get up and stop all the bowing; it was OK and let's just forget it and finish dinner.

A few minutes passed and I thought it would be a good time for me to ask the captain a question, so I said to the interpreter, "Ask the captain if I might ask him a question, if you don't mind." "What is that?" asked the interpreter. "Well, I'd like to know why the captain thinks Japan lost the war."

The interpreter translated the question for the captain and he just looked me right in the eye and very calmly told me, "Japan lost the war because we underestimated the technological capability of America, but that won't happen again the next time!"

That's exactly what that guy said to me; verbatim; word for word. I remember it to this day. I would tell people about that over the years and they would just look at me like they didn't believe me, but it is the truth. The Japanese didn't really lose the war if you look around at what's happened since. You always hear that history repeats itself; we never really learn from the past. I've lived long enough to see that what was happening before the war is happening again today. It's crazy.

That Japanese captain gave me a medal of his later on when he was processed, just before being sent back to Japan himself. It looks like it's pewter or silver with a gold star on the front of it and has a bunch of Japanese writing on the back. I don't know what the writing says since I don't read Japanese. I was told , though, that he earned it for something he did in combat; bravery in combat, perhaps?

If I had ever seen a medal like that on the uniform of a dead Jap, I would have pulled it off and kept it. You know what I mean? That Jap captain gave it to me, though, and I never even asked him for it. He gave it to me freely; of his own free will. I still have it to this day.

What is particularly noteworthy in our Marine Sergeant's story is that it seems obvious history does have a tendency to repeat itself when one considers the root causes of many of our wars. Historians will agree that misunderstandings have been at the core issue of some of history's most terrible conflicts. In his recent book, Yamamoto, Reluctant Admiral, published by Kodansha International, Ltd., the author, Hiroyuki Agawa quotes Admiral Inoue Shigeyoshi, onetime head of the Japanese Naval Affairs Bureau as saying shortly after the war,

> Japanese politicians and Army men both underestimated America's natural strength and the spiritual strength of its people, particularly its women. They had a childish notion that, since women had such a powerful say there, it wouldn't be long before they started objecting to the war....
>
> At the same time, it is the view of many, that the American people also underestimated the strength of Japan as a foe to be reckoned with. Prior to the war, a majority of Americans, had in fact, accepted the view of Japanese as that depicted in cartoons and racist drawings of the time. To Americans as a whole, the Japanese were nothing more than funny little men with buck teeth and round, ill fitting spectacles, who were physically incapable of flying airplanes because of the structure of their eyes.
>
> Regardless, one can say with certainty today that rampant ignorance and contempt on opposite shores of the Pacific played to the disadvantage of both peoples when one considers the millions it ultimately affected.

William Schultz, assigned to the 10th Army at that time and stationed in Noumea, New Caledonia, had loaded up Army ships for the invasion of Japan. Though he had spent 19 months on the New Hebrides battling the *Anopheles* mosquito, he had no points accumulated whatsoever. Most of the members of his unit CMB538, however, did, and were promptly returned home to the States. Schultz was given the choice of escorting Japanese prisoners to Japan on flat bottomed LSTs or staying on in Noumea to unload the 10th Army invasion ships and destroy now surplus equipment and supplies contained therein. Not thoroughly convinced of the safety or comfort of an LST, Schultz chose the latter and started post haste to smash, crush, and otherwise make unfit for use everything the Army owned, from huge radios to jeeps, trucks, bulldozers, and earth movers.

William K. Schultz CM3/c, USN Malaria Control CMB538 SeaBees. Noumea, New Caledonia, was the first city I had seen in over nineteen months, so it was a real treat. There must have been thirty or more bars there, all filled with service people. In the center of town, there was a Government Beer Hall with a fence around it. It had picnic tables and no roof. I remember there was a small hotel there where I ate steak and eggs a few times, which was kind of a treat since all of the eggs we had had for the past nineteen months were of the powdered variety and did not taste very good.

Most of the city of Noumea had no running water and no sewer system. They had what we called, the "honey wagon," which picked up and emptied the chamber pots set out in the streets. It was no pleasure to ride behind this truck, believe me. Also, the people there would urinate in the gutters in front of their homes, both men and women.

I remember that the island had many small towns and even a large deer population, some of which were hunted by the locals for their hides. Noumea also had a few nice beaches, but not as nice as Santo and did not have the jungle the New Hebrides did, so in that sense, it didn't seem at all like a South Pacific island as we were used to.

Noumea also had a very large nickel mine which I think was owned by Hoover. The natives there were called Kanakas. Some of them worked at that nickel mine and some of them helped to unload all the ships in the harbor.

There were only a few small grocery stores there that I remember and they would deliver fresh bread around the city, depositing the loaves that had no wrappers, into boxes much like our newspapers were back home.

There were probably 100,000 or more service people in Noumea at the end of the war, since the 10th Army was being assembled there for the invasion of Japan. My outfit was attached to the 10th down there. Most of the ships that were to carry us to Japan were already loaded and ready to leave when the word came in of the Japanese surrender. It was now our job to unload these ships, which took about three months. After we got the ships unloaded, we destroyed everything.

Brian Taylor, PhD, Nottingham, UK, Onetime Government Entomologist Malaria Eradication Programme, Solomon Islands. Listening to this reminds me of the story, unconfirmed, that, while on Guadalcanal after hostilities there had ceased, the U.S. military authorities offered to build a decent road right around the whole island. That would be a useful training exercise for the U.S. engineers and would utilize equipment already on the island. The just returned British administration, with their still unchanged limited foresight and wisdom, saw no merit in such a road, presumably thinking only of the future maintenance costs and, no doubt, having the view that the roads on north Guadalcanal served the coconut plantations. Elsewhere on the island there were only local villages and, thus, no use for roads! To this day the island does not have a circum-island road.

Bill McLaughlin MSgt., 221st Field Artillery battalion, Americal Division. When I was discharged, I was given an automatic 10% disability for malaria but, after one year with no attacks, the disability was reduced to 0%. I had malaria twice out there, dengue (breakbone) fever, jaundice, and jungle rot as we called it. Both my eardrums were blown out by muzzle bursts at different times and I got shrapnel in my left arm. I was hit with a ricocheted rifle ball which burned my leg. I also hacked my leg doing malaria control and had six stitches in the tendon with five in the skin just outside my right knee. But for those three and a half years, I coped, finally rising in the Recon Troop to Scout Sergeant and then finally to Platoon Sergeant.

Robert Otto, BM3/c USN, USS Grayson. I had just been discharged from the Seattle Naval Hospital a few days before V-J Day and was in Los Angeles. I was not in the best of shape then and if memory serves, I probably did not care one way or the other about V-J Day.

Robert Michel PhM1/c, USN Malaria Control Unit Cactus. I do want to tell you about Dave and I, though. We were together and then apart and then together, all through the war. After Dave was discharged, Dave and I lived off the 52/20 Club. The 52/20 Club, as it was called, was a program for the veterans. I am a little hazy on the details, but I recall that for 52 weeks, the government paid us $20.00 per week if we didn't have a job or were not going to school. Dave and I rented a walk up two room apartment on the second floor on West 8th in the Village and had the time of our lives during the summer of 1945.

Dave had an aunt who also lived in the Village and worked for some kind of advertising company and was able to supply us with all the Pinch Bottle Haig & Haig we wanted. At that time Scotch was in very short supply in the United States and to be able to get Pinch Bottle was something else. We would sit in our scivvies with our bare feet propped up on the sill of the French window with a glass of scotch in our hands and a life-sized frontal view painting of a nude brushing her long hair hanging on the wall behind us. I think we paid twelve dollars for the painting.

Artistically it was probably not very good, but anatomically it was superb. When the light was right, people walking along the other side of the street below could see not only Dave and I in our bare feet with the glass of scotch, but also the nude behind us. We saw one heck of a lot of tourists snapping pictures of us. Like I said, it was one helluva summer. But then we decided it was time to go back into the real world, and we went our separate ways.

Chapter 12

The Letter from Nimitz

FATE HAS ALWAYS been credited by we mere mortals with deciding the outcome of those events we either do not understand or have no control over. Interestingly enough, the letter from Nimitz was one of those strange quirks of fate. It was interjected into the lives of an unknowing few, serving in the furthest reaches of the South Pacific in the spring of 1944, twisting and turning as it played its hand. A fate so fickle that when it dealt the last card to Bob and Dave, the winning hand they were left holding just probably saved their lives.

In one respect, the story told by Bob and Dave of the Skeeter Beaters is a straight forward one. Quite simply, it is Bob and Dave's take on their experiences and how they remember them today, 58 years after they occurred. Within their story, however, lies another story that surfaced in their telling, and that is the story of malaria control itself. A story hidden in a convoluted series of seemingly unrelated events. A story, when viewed in the larger context, paints a most remarkable picture.

Having read the stories thus far, the reader is left to accept or reject them. Bob, Dave, and many other sources (whether personal recollections, official records, or literature) were generous in their presentation of individual events that contributed to the often confounding story of malaria control. It is only by viewing the individual puzzle pieces in total, however, through the magnifying glass of 58 years hindsight, that the beauty of the entire landscape slides into focus. What we are left with is the beauty of history itself—the ordinary and remarkable circumstances of fate that has today become a part of our patrimony.

The story of malaria control is woven throughout the narrative by Bob, Dave, Dr. Arthur King, the Official History of the Navy Medical Corps, our Marine Sergeant, William Ricciardi, Robert Otto, Bill Dickman, Tom Croft, and amazingly enough, the War Department itself. All of these provide integral pieces; yet if viewed individually, none can possibly convey the significance of malaria control to the outcome of the war in the Pacific. By carefully analyzing the questions raised by events described throughout the narrative, the seemingly unrelated puzzle pieces start to take shape.

For example, how was it possible for LtCmdr. James Sapero to openly solicit, and take, volunteers for his outfit when men and supplies were at a premium? Why was Malaria Control Unit Cactus flown from Efate to Henderson Field—using valuable troops and military equipment—when the situation was so desperate? As William McLaughlin related, he and his unit of the Americal Division were "thrown in the path of the Japanese like a log in the road" in the spring of 1942. The United States faced a desperate situation at that time in the South Pacific. Also, the Marines had only landed on Guadalcanal August 7th. The few troops ashore at this point were in terrible need of all manner of supplies, including the most essential—food and ammunition.

How could a small malaria control unit make the incredibly dangerous flight to the island, much less take up critical cargo space that could have been used to carry in more critically needed supplies? How could it have been possible that a Marine DC3 was filled with mosquito nets and flown into Henderson Field during one of the first supply missions flown by SCAT? Fifty-eight years after Bill Dickman's first flight into Cactus on a supply mission with MAG 25, he examined his flight logs and verified this. So we are faced with a story that seems somewhat out of context when examined in light of other accounts of the war in the South Pacific at this time. Yet, a plane load of mosquito nets not only preceded many critically needed supplies, but more than likely displaced items valuable to the battle such as aviation fuel, bombs, ammunition, and food

Bob's story of receiving the first known " bug bomb" is another example. Again, an odd item to show up at such a crucial time in the siege of Guadalcanal—mid-September or early October. Once again, valuable cargo space was made available for the delivery of equipment to MCU Cactus—a crate of experimental aerosol insecticide dispensers. Who determined the bug bomb was more valuable than food?

How was it that Bob and Dave, mere enlisted men, were granted a 30 day R&R on New Zealand, right in the middle of hostilities in the South Pacific?

Why did Arthur King and others express such strong animosity towards LtCmdr. James Sapero, at times almost bordering on hatred?

Why did it take the U.S. Army almost two full years to take over malaria control efforts in the South Pacific, from the spring of 1942 to the fall of 1944?

Is this just a crazy story that is part fact, part fiction, and part memory lapse after 58 years? We have seen a number of clues throughout the narrative, but let's look a little closer at the facts.

Early on Bob Michel stated, "I do recall one of our officers saying that we had a letter from Admiral Nimitz that gave us top priority to fly anywhere in the South Pacific."

Later, we were given another integral piece of the puzzle. Dr. Arthur King (U.S. Army Surgeon, Task Force 6814 and later chief medical officer for Army Service and Supply South Pacific) mentioned a high ranking Navy Malaria Control officer, attached directly to the staff of Admiral Halsey (previously Admiral Ghormley), ComSoPac, who made it difficult, at best, to cooperate with the Navy. During the summer and fall of 1942, virtually all of the planning and implementation of malaria control initiatives in the South Pacific was controlled by the Navy. It was the Navy who, shorthanded, expected the Army to execute the plans and supply all the necessary medication, tools, and manpower.

Another critically important piece of the story of malaria control was found deep in the History of Navy Entomology. Compiled by Commanders Peterson and Wooster and later updated by LtCmdr. Need, we are told of the devastating effects malaria was having on troops stationed on the New Hebrides during the spring and summer of 1942. Their record of events details for us the fact that Ensign Kenneth Knight led the first Navy survey and control teams dispatched to Efate, arriving on August 12th. This is a key date when we consider Bob's story of arriving at Efate, staying only a day before flying out for Guadalcanal and arriving there on approximately the 13th.

Finally, we heard Dr. Arthur King describe the bitter rivalry between the Army Services and Supply group and the Navy malaria control units which finally reached critical mass on September 10, 1944. When the War Department issued Circular #178, the order effectively ended the Navy's command of malaria control efforts in the South Pacific and placed that branch and all its resources under the direct authority of the Army Service and Supply group.

When viewed in total, these facts provide both confirmation and continuity to Bob's recollections. More importantly, they open a door to the oft-slighted history of malaria control and the impact it had on the war effort in the South Pacific. LtCmdr. James J. Sapero, MC, was in fact the commanding officer referred to as the high ranking naval officer who was placed directly on the staff of Admiral Ghormley ComSoPac, and later Admiral Halsey, when he assumed command (Ghormley was relieved in October 1944). It was also Sapero who was referred to as the high ranking naval officer connected all the way back to Washington.

It was in fact LtCmdr. James J. Sapero who was described as the commanding officer of the malaria control unit in the South Pacific in the Navy's final report. This report was compiled shortly after the war and detailed the unit's activities. In it, Sapero himself tells us of those first days. "This landing [Guadalcanal] under combat conditions was followed by repetition of many of the same mistakes which were made on malarious [sic] islands under noncombat [the New Hebrides] conditions."

Now, 58 years after the events, we know that at the outset of the war, James Sapero was given a mission. The orders for that mission emanated from the highest authority in the War Department and the Department of the Navy. That mission, quite simply, directed him immediately to the South Pacific. He was to do everything and anything within his means to protect the thousands of U.S. troops who would soon be exposed to diseases rampant in the area that infected the locals on a wholesale basis—most specifically malaria.

In other words, Bob was right—there was a letter from Nimitz. Was this letter literal? While I was unable to unearth a copy of the actual written document in my research, is it possible that a letter did or still does exist? Perhaps. What is significant to Bob's story of the Skeeter Beaters, however, is the figurative reality of the letter—the absolute implicit authority for Sapero to complete his mission in controlling malaria during the war in the Pacific.

Sapero was a man who had no chain of command to answer to—not Ghormley, the Navy's South Pacific commander; not

Patch, the Army general in command of the South Pacific. And believe it or not—not even Nimitz. A relatively low ranking LtCmdr., Sapero owed his allegiance to no one except those in the highest levels of the Department of the Navy and the War Department in Washington.

Consider the long reach of Sapero's authority. How is it that a plane load of untrained enlisted men, led by an entomologist and medical doctor, could be flown into Henderson Field so shortly after the 12th of August? Sending his malaria control personnel up to Efate under the nominal leadership of Kenneth Knight, he split that first group into two units, leaving Knight and half the men with him on the New Hebrides. Sapero then pulled rank, so to speak, and commandeered the needed plane and crew to fly the other half of his men, with another Navy entomologist by the name of Robert Mount to Guadalcanal. No problem for a man who had the muscle of Washington to back up his orders.

Given a mission to perform, Sapero didn't stop by just sending a few men to the 1st Marine Division on the Canal. Whenever he had it available, he followed up with equipment he thought could be put to good use. Witness the delivery of mosquito nets flown into the Canal on September 1st. Sapero had the mosquito nets, and it was his opinion that they were important enough to get them to the Canal immediately so that they could be put to good use by troops exposed on the malaria-infested island. He commandeered space on the earliest available SCAT flight loading up on the New Hebrides and, once again, waved the letter, and the mosquito nets were on the way. Later in the month, he had experimental aerosol insecticide dispensers at his disposal. Here as well, a mere mention of the letter and a crate of them is made available to the men of MCU Cactus.

In December, when military command on the Canal was transferred to the Army and the 1st Marine Division was effectively rotated out over the next six weeks, the men of Malaria Control Unit Cactus were left behind. Little wonder MCU Cactus didn't head down to Australia with them when we recall what Bob told us: "Even the 1st Marine Division brass didn't know what to do with us." With the Army takeover complete in January, the men of the orphan unit MCU Cactus were in jeopardy of falling under the control of General Patch and the Army (remember War Department Circular #178) on the Canal. Sapero once again stepped in and yanked his men off the island and spread them out over Tulagi and the Floridas, ordering them to continue their survey, control, and eradication duties in an area where they would remain under the Navy's (and ultimately his) control.

By June 1943, having served in the South Pacific for a year now, Sapero felt the men of Cactus needed a break. He decided a little R&R would be in order. To get Bob and Dave their well deserved rest, Sapero simply pulled out the letter and forthwith had them transferred to the ship's company of the Naval Hospital in Aukland, New Zealand. By transferring them to a new duty station, Sapero was able to deliver the break his men needed from the rigors of the South Pacific campaigns, ensuring there would be no questions asked when two enlisted men showed up in New Zealand for one of the biggest parties of their lives.

Through the better part of two years, Sapero was continually at odds with his Army counterparts. Arthur King, MD, who sat on the staff of General Patch, is to this day most critical of both Sapero's knowledge of the disease of malaria and his methods of dealing with it.

Their final showdown took place on the New Hebrides in the fall of 1944. Sapero ordered the Army Service and Supply command on the island to transport what he considered to be an infected native population, away from U.S. military installations by dusk of each day, interring them in isolated camps until dawn. The only problem with his plan was that he included all natives in the vicinity of U.S. installations, even those employed by the island's most influential French plantation owners. Upon hearing that they would be without their personal servants, and deprived of many of their best farm workers during the hours of darkness, the French ran to the Army and Dr. King to intercede. The battle of Armageddon was on. King went up his chain of command only to hit a brick wall—General Patch, the Army area commander would not contravene Sapero's orders. King next took his argument to the Navy, where it ended up on the desk of Admiral Bull Halsey, overall military commander of the South Pacific. Halsey would not intervene in the squabble, nor would he order Sapero to back off. Next stop was the desk of Admiral Nimitz, Navy commander of the entire Pacific theater. Again, no luck for Dr. King or the Army—Nimitz would have no part of the feud. Sapero was finally thwarted in his efforts when the French High Commissioner in Espiritu Santo made a personal plea for relief directly to the United States State Department in Washington, D.C. It was then, and only then, that Sapero's authority finally reached its limit. On September 10, 1944, the War Department issued Circular #178, countermanding Sapero's orders on the New Hebrides. Command of all malaria control efforts were turned over to Dr. King and the Army Service and Supply Group.

The story of the Navy's last stand against the Army over the ultimate control of malaria efforts in the South Pacific tells an interesting story about how Sapero's peers in the military viewed him then and after the war. Sapero disliked by many and criticized soundly by medical professionals for his efforts to locate, eradicate, and treat the deadly onslaught of malaria. He has been impugned professionally and personally and, it should be noted, has admitted himself to mistakes in carrying out his mission. But is that all there is to it?

Robert Michel, PhM3/c USN Malaria Control Unit Cactus. I agree that Sapero was probably disliked, but I think it would probably have been for his position, not for any personal reason regarding him as a man. All I can tell you is that when Dave and I met him on the beach in Noumea, I remember him as a very pleasant man. He was unassuming, polite, and treated Dave and I with respect. He was just a regular guy to us—he had no airs about him that you might expect from a career naval officer.

Again, hindsight lets me take a broader view of the situation. Had James J. Sapero been Albert Schweitzer himself, I am convinced he would have been just as maligned. Remember, Sapero was a man who had been singled out, not only to perform a mission critical to the war effort, but was a man given every privilege along the way. Traveling in the loftiest circles of the military elite, Sapero was equal in power to the highest ranking admirals and generals he came in contact with. Sapero was also a man well-suited to the mission, and a man who knew well how to wield his power. He was able to order the highest ranking Army officers to do his bidding when moving his odd mix of men and equipment, quasi-military units that were considered extremely low priority to the fighting commanders, who placed more faith in bullets than mosquito nets. Is it any wonder then that Sapero served as a lightning rod for all of the envy, animosity, ill will, and power-brokering that took place in the South Pacific?

Regardless, it was James Sapero who did in fact succeed in performing his mission in the long run. When the war started in 1941, Sapero, like every other military commander given a job to do quickly, found himself understaffed, with troops woefully under-trained and ill-equipped. Arriving in New Caledonia in June 1942, Sapero had just six weeks to staff and equip his units and deploy them in the areas he deemed most needed. He did what he had to do to get men like Bob and Dave to volunteer, and he did what he had to do to steal them from units like MOB5. He accepted nothing less than absolute obedience from everyone he came in contact with when he commandeered the planes he needed to move his men and supplies. He allowed no man to stand in his way, regardless of their rank.

Starting with a nucleus of just one entomologist, a single medical doctor, and less than 40 men in 1942, the dedication of James Sapero and his men resulted in the ranks of Navy malaria control growing to over 9,000 men by the end of the war in 1945. Over 200 entomologists and 122 malaria control units were strategically situated throughout the Pacific wherever American troops were stationed and more importantly, where malaria was no longer a contributing factor to the daily casualty reports.

As heavy-handed as many who remember him would like us to believe, and as crazy as others felt his methods, Sapero never forgot his men. In the spring of 1944, when supplies and replacements were finally pouring into the theater, Sapero selected Bob and Dave to participate in a newly instituted rotation program for men of the Navy and Marines who had served in the South Pacific—one that had far reaching consequences on their lives. Pulling out the letter from Nimitz in May of 1944, Sapero had Bob and Dave transferred. This time it was back to Noumea, New Caledonia, and their original unit, MOB5, where they were slated to stand at the head of the line for a one-way trip home to the States. After the New Georgia and Cape Gloucester campaigns, and manpower levels from new replacements were considered adequate, the Navy and Marine brass were resigned to a long, drawn-out war with many bitter campaigns to come. Strategic thinkers came up with a way to reward the battle-weary by rotating them to duty stations in the States, while at the same time seeding the ranks of new units being formed for future campaigns with battle tested veterans. While very selective in its application, yet maintaining a sense of fairness to all, a point system was instituted.

Marine Sergeant, Section Leader 75mm Pack Howitzer, I/3/11. After Cape Gloucester, we went straight to Pavuvu to get ready for our next landing. We heard it was going to be at a place called Peleliu, though we didn't know where that was. Roughly in May 1944, they came up with a point system for guys who had at least two campaigns, which I did—Guadalcanal and Cape Gloucester. You got two points for each campaign and one point for each month overseas, that's it. So, I had 27 points—that's what you had to have. In June, I sailed on the USS Neville right there from Pavuvu and went straight to San Diego. No stops.

It's a good thing I made it back, too. I probably wouldn't have survived Peleliu. Nineteen guys I served with were all hit in their landing craft going in to the beach by a Jap 90mm mortar. None of them made it. I would have been right there with them had I not gotten out on those points.

Bill McLaughlin, PFC, 221st Field Artillery Battalion, Americal Division. I remember the point system came out in 1944 when we were still on Bougainville. It was so slow and tedious. At most 2 men per company, battery, or troop were selected. So, very few of us paid any attention to it. It seems to me the points in the Army setup were worth more like 5 points for a campaign.

I got busted when I was with the Field Artillery—a court martial for taking a green and obnoxious 2nd lieutenant down a peg. The language looked awful in print, but that was the way we talked back then. Anyway, I lost some points due to that court martial, but gained them back later with the decoration of the Silver Star. At the end of the war I had more than 100, but can't really remember how many.

I ended up coming home in 1945 with the other old men who had been the originals out there. We had shipped out of Brooklyn, New York, some six weeks after Pearl Harbor. The war finally ended when we were off Hawaii on the way home. I remember we had a party on Cebu (the Phillipines) the night before we left and a new guy said to me, "Gee, Pop, it's a shame they leave old guys like you over here." I said, "That's right, Sonny. How old are you?" He said he was 26. I was a year younger than he was.

My Lieutenant, a Platoon Leader who had gotten a field commission on Bohol, Philippines (I took his place as Platoon

Sergeant), came down that night to ask me if I'd go with him to the newly formed Philippine Scouts. It was like the Navy Seals, going in a few days in advance of the rest of the troops and demining explosives. He was utterly fearless. I said, "Maury, I've been over here 3-1/2 years. I'm going home."

I was tempted, though. He went on to Japan for the occupation.

John Francis Richter, PhM3/c Corpsman USN, 75mm Gun Half-Track, Headquarters & Service Battery, 11th Marines. I am a complete blank on the point system! I must have gotten back to the States just before the institution of the system. When I got there, I had my hands and thoughts full, focused upon adjusting to my new Stateside duty, plus fighting off recurring attacks of malaria!

Later, I was saved from further overseas combat duty—and of my possible demise—by those "Angels of Mercy," the A-Bombers.

Robert Otto, BM3 USN, 20mm Gunner USS Grayson. I knew or heard about the point system but I did not pay too much attention to it. I knew at the time that I would have more than enough points, though I am not sure how they awarded them. I was not interested in leaving the Navy anyway, I wanted to stay in.

Later on, the doctors at the Seattle Naval Hospital said I could not pass any type of physical exam. So ended my career.

Dave Minton, PhM3/c USN Malaria Control Unit Cactus. Bob and I came home on points. There was a point system and we had enough. Bob may not remember that, but that's what happened. Now how we were selected for the point system, I couldn't tell you. There must have been some selection process but I don't know what it was.

Shortly after their return to Noumea in May 1944, when the opportunity presented itself, Bob and Dave found themselves assigned to the ship's company aboard a homeward bound troop transport—happy to be returning home, regardless of how they were selected. Ironic isn't it, when we remember Dave's comment about joining the Navy three years earlier in 1941, to be a sailor. When Bob and Dave finally became sailors, it was at a time and place in the war when men and supplies were moving forward into the teeth of stiffening Japanese resistance at places we're all familiar with—Saipan, Guam, the Philippines, Iwo Jima, and Okinawa. By selecting them to participate in the point system rotation, Sapero had just rewarded two of his first volunteers with a well deserved early return to the States. He made sure Bob and Dave were on that troopship headed home accompanied by their unit commander, Kenneth Knight.

Who knows—perhaps no one ever will know for sure, but it just may have been the letter from Nimitz that saved the lives of Bob and Dave.

Chapter 13

Life Is Good

September 1945-December 2000

Robert Michel, PhM1/c, USN Malaria Control Unit Cactus. When I first got out of the service, I worked as a guard at a Bendix machine shop in Teterboro, New Jersey, for a few weeks. Then after Dave and I decided to go back into the real world, I went over to New York City and paid a visit to Jack Costello who, you will remember, was an announcer at NBC. He suggested I go to work for him there. As a result of that meeting with Jack, I landed a job as a Page at NBC in New York City for about six months. There were a lot of young guys and girls working there as Pages while they were going to school. One of the people I met on the job at NBC was the girl that Dave later married. I had actually introduced her to him—at his urging—one day when he stopped in.

Another person I met while working there, whom I became good friends with, was a guy by the name of Bob Keeshan. He would occasionally come over to my house in Hackensack on weekends and on Thanksgiving for some home-cooked meals. We had both attended an announcing school run by the chief announcer at NBC, a man by the name of Pat Kelly. Pat had me at the head of the class and was going to make me a Junior Announcer at NBC. Bob Keeshan was at the bottom of the class. Unfortunately, at that time, the union, AFRA (American Federation of Radio Announcers) was negotiating a new contract with the networks and one of the provisions agreed upon was to eliminate the Junior Announcers. No more Junior Announcers. NBC was told to hire the announcers they already had or to fire them, but to add no new announcers. So, I didn't get that job.

I got my first commercial radio job at WHIT in New Bern, North Carolina, with the assistance of Pat Kelly. From there, I moved around the country and ended up in Altoona, Pennsylvania. I met my wife there, Marj Good, at a party in 1947. Five months later we were married, and we're still together 53 years later. When we were on our honeymoon we stopped in New York City to meet with Dave and his wife, Loraine, at a little Italian restaurant down in the Village where Dave and I used to go. We had a nice time together, but then it was all too soon, time to leave.

It was some time later that Dave, who loved practical jokes, came out to Altoona, Pennsylvania, where I was working. He brought his wife and Muriel Scharf, an old girlfriend of mine from New York. Needless to say, Marj was fit to be tied that Dave would bring one of my old girlfriends. But we took them to some of the night spots and a restaurant named Erculeanis's (it became quite famous some years later and relocated to NY). That was 1948 and the last I saw Dave for about forty years.

I visited Dave in the Keys, where he now lives, a couple years ago. In reminiscing, I learned that Muriel Scharf and Dave's wife remained good friends, and that Dave had received a letter from Muriel when his wife Loraine had died. I persuaded Dave to call Muriel in Canada, where she lives as a widow, and surprise her.

After she answered the phone, Dave told her there was an old friend who wanted to say hello. I took the phone and told her who I was. The first thing she asked was, "Are you married?" When I had been going with her years before, I would be invited to Sunday dinner at her house where she, her mother and Grandmother lived. Her Grandmother made the best mashed potatoes I have ever eaten. She used at least a half pound of butter and about a half can of evaporated milk. I have tried to duplicate the recipe ever since, but it never turns out as good as it did then.

Later, after Altoona, I worked in Johnstown, Pennsylvania, for a short while. I moved on to Cortland, New York, and then to Syracuse, where I worked at radio station WSYR.

In 1950, the Navy tapped me on the shoulder again. I was called back into the service during the Korean fracas. I was again stationed at St. Albans Hospital on Long Island.

While I was there, just before being discharged, I went over to New York City to stop by NBC, and who should I run into but my old friend, Bob Keeshan. He had been playing the part of Clarabell the clown on the Howdy Doody Show and was looking for a guy to do a pilot for a kids' TV show he was putting together. He wanted to produce it and asked me if I would be the central character. When I asked him how much it paid, he just laughed. He told me it would be a pilot on speculation and that no money was involved. I told him I had a mortgage, a wife, and a three-year-old daughter and that I needed an income. As a result, I had to turn down the offer. The pilot was picked up and ended up being Captain Kangaroo. I could have been Captain Kangaroo. It's strange how fate works.

After receiving my discharge again from the Navy, I went back to Syracuse and got a job at W H E N television. After a couple years there I got a job in 1954 at the Dumont TV station in Pittsburgh, doing a half-hour show on Saturdays. For an entire winter I commuted back and forth from Syracuse, New York, to Pittsburgh by train: the New York Central to Buffalo, then the

P&LE to Pittsburgh. The cars were old with seats that did not recline, and there was very little heat in them during the all-night trip. It was cold. After that first trip my wife bought me a Harris Tweed suit. The material was so thick it couldn't be pressed, but at least I stayed warm. The pay for the show I did, which was sponsored by Mars Candy Bars, was $50.00. My expenses were $45.00 for a net profit of $5.00, but that show got my foot in the door in Pittsburgh. At the time, this was my only income. I would do a live opening, a closing, and three one-minute live commercials inside the show selling Mars Bars to kids. I was dressed as a cowboy sitting on a rain fence at the C-Bar Ranch. The cameramen and crew knew I was hard up for money, so they would give me cartons of candy bars, the ones we used as props, to take back home with me. We put them in the freezer of our refrigerator and sometimes they were our lunch or supper.

I eventually went full time at KDKA radio in Pittsburgh on the all-night show, as well as being the early afternoon disk jockey. I also did a lot of freelance at the same time. I got into the Screen Actor's Guild (SAG) from a movie, "The Fish That Saved Pittsburgh." I ended up on the cutting room floor, but it got me into SAG. I did a lot of small parts in many movies shot around here and finally retired about three or four years ago.

I also own a motorcycle business that I started in 1958 as a hedge against further unemployment. It has grown under my son's direction, Greg, to about 80 employees and one hell of a payroll. We also have a web site now (partswarehouse.com and worldofcycles.com) and an 800 phone number for parts sales all over the world.

Over the years, I have been able to contact only one of the group of Pharmacists Mates I landed on the Canal with, Dave Minton.

And here is another story about Dave. Dave reenlisted in the Navy in 1950. When he volunteered, he figured that as a Pharmacist Mate First Class, the worst that could happen would be that he would be assigned to a rear area medical facility. Instead, they sent him to combat training at Parris Island, and then to Korea as a Company Medic. He ended up again right back with the 1st Marine Division, only this time at the Chosin Reservoir. That's Dave's story, though.

Dave Minton, PhM1/c USN Company Corpsman, 1st Marine Division. Remember, everything I ever did in the Navy, I volunteered for. From the time I first signed up in 1941, to Malaria Control in the South Pacific, to Korea; everything I ever did was by my own choice. I have no regrets about anything that happened to me because it was all my own doing.

Robert Michel, PhM1/c USN Malaria Control Unit Cactus. I had tried for years to find Dave, and then one day a few years ago, I located a Dave Minton from Connecticut on the Internet. I phoned him and was told he was not the guy I was looking for, but that I should try a number in another area of Connecticut. I called the number and got a woman on the phone who said she was his daughter-in-law and gave me a number in the Keys. Now, I am told I have a distinctive voice. I called that number and when a male voice answered, I asked, "Dave Minton?" "Bob Michel!" was the reply. After all those years, he remembered my voice. A few months later, he stopped by on his way to Connecticut to sell his house up there. How's that for a story!

Chapter 14

Looking Back

August 2000-February 2001

GUADALCANAL AND THE Florida Islands, made up of Tulagi, Gavutu, Tananbogo, Nggela Sule, and Nggela Pile, were all part of the British Solomon Islands in what was called the lower Solomon chain. Our stories have been limited to the most historically significant islands of the Solomon Islands. The entire chain, however, consists of an unbelievable 992 islands of which only a paltry 300 or so are inhabited.

Guadalcanal is ninety miles long and thirty miles wide, while Tulagi is a dot on the map in comparison. Viewed from the air or from the sea, the islands present a breathtaking view. High mountains rising up from the beaches reflect brilliant sunshine, glittering off effervescent blue waters and broken by colorful coral reefs. Lush vegetation abounds. With stately coconut trees lining the shores, the islands present visitors a picture-postcard of awesome natural beauty. It is difficult to imagine that a paradise so peaceful from a distance could have been the site of some of the most vicious battles of World War II.

Brian Taylor, PhD, Nottingham, UK, Onetime Government Entomologist Malaria Eradication Programme, Solomon Islands. In some ways, the anti-malaria operations used in the Solomon Islands in the 1970's differed very little from those of the war years, but we did no aerial spraying or any mechanical operations such as ditch digging. had long been given up by this time, but chloroquine was used as a prophylactic by most expatriates and tourists. It still is, in fact, even though it is not always that effective.

In Honiara, the capital city of Guadalcanal, and in other towns, some of the ditches and pools were still oiled; not to much effect, though, since mosquitoes fly. The main technique we used was the spraying of inside walls and under the eaves of all the houses. That was done with DDT, a very good and safe insecticide for that purpose, as well as being inexpensive. A case detection system was put into place, and those with positive blood slides were given treatment.

Nowadays, the main antimalarial thrust is the use of pyrethin-impregnated mosquito nets, which do have an impact on malaria levels in young children. However, as I later proved, the main vectors bite out of doors in the very early evening hours when humans are not under mosquito nets.

Our programme in the Solomons was very successful, but poor management led to the situation reverting to preoperational malaria levels within a short time after my contract ended there.

Robert Michel, PhM1/c USN Malaria Control Unit Cactus. During the years that followed, I have heard that the South Pacific Islands are a paradise and some of the most beautiful places on earth. Believe me, at that time during the war, they were far from beautiful. I saw the destroyed areas—palm trees cut off at the tops from bombs and shells. Decaying bodies were strewn all around with mud, rain, and disease. I have read articles about Guadalcanal that give glowing reports, even though malaria has not been completely eradicated.

The water, though, that was altogether something else. While in the Florida Islands, I used to go pearl diving with the natives, or sometimes scuba diving along the waterway between the Big and Little Florida Islands. It was spectacular. The bright colored coral and the tropical fish were fantastic. I got a small scratch on my leg from the coral one day, however, and got blood poisoning from it. I was in big trouble for a while. So, even in the beautiful places, it could be dangerous if you weren't careful.

Locals throughout the South Pacific have always been aware of the dangers inherent in a coral cut and are quick to instruct a victim to douse the wound immediately with fresh lime juice. The acid in the fresh lime will kill the living animals embedded in the coral. If not treated immediately, those same animals can thrive and grow under a victim's skin, causing a really nasty infection.

Tom Croft, BM3/c USN, Gun Loader Turret-1, USS Monssen DD436. It has been almost 59 years and I can still see Guadalcanal and that little Savo Island. Lots of ships were sunk around that place.

PFC Stewart Winterton, Marine Rifleman, B/1/7. It's been over 50 years and I only recently started eating rice again—couldn't eat it for years. I got sick of it.

Marine Sergeant, Section Leader 75mm Pack Howitzers, I/3/11. I joined up with one outfit in my life and I'll never join another—not even the Boy Scouts. I never joined any of these veterans' organizations or associations. That was enough for me. I heard the First Marine Division Association was taking groups of guys back to the South Pacific over the years. I would never go back—that's not for me. I hear now the natives down there are even going after outsiders and have some sort of revolution going on. We really live in a strange world.

I have to say, after the first twenty five years of my life, God has been good to me. I have had a great life and everything has gone well for me.

William C. Dickman, MSgt. USMC Enlisted Navigator/Naval Aviation Observer VMJ-253 MAG-25, South Pacific Combat Air Transport, SCAT. I have been back to the Pacific once since I retired from the Marine Corps, and I would like to return again once more before I pass on.

Williiam Ricciardi, MM2/c Boat Gang USS Crescent City. No. I never really wanted to go back to any of the places in the South Pacific or the Western Pacific. I saw too many bodies out there. I saw hundreds and hundreds of bodies and parts of bodies—most of them were naked. Off Saipan, the last island I hit, I saw too many empty life jackets in the water. I'm afraid that if I were to go back there now it would dredge up too many bad memories. It's a terrible thing to have to remember. You know, if you have ever seen any combat movies, I mean real combat movies, when a guy kills another man, the first thing he does is get out of there. He doesn't want to stay around. He doesn't want to look into the face of the man he has killed because he can't forget that. It's a look of shock, just like a baby's face—they know they are going to die and it's a terrible thing to see.

Bill McLaughlin, PFC 221st Field Artillery Battalion, Americal Division. You asked me if there was any place I'd like to see again where I served during the war. First off, I thought, No. But then, perhaps I would like to see Fiji and Australia. None of the areas where we fought, though.

I wrote a manuscript on my year in the Recon Troop back about 30 years ago when our first grandchild was born. I thought he might someday wonder about what I did. He never did wonder, but it gave me a chance to run copies on the typewriter and send three of them to men in my platoon who were involved in the actions. The details were still fresh in my mind then, and I asked my buddies to correct anything amiss, or supply facets to the fights we were in. None of them corrected anything. Some did, however, supply names of others who were with us. One is in a nursing home, one is dead, and the third is now in a nursing home unable to communicate after all the shocks he sustained. My former scout sergeant of the Recon Platoon, whom I recently contacted of late, can't remember much of anything about those days at all now.

Time is going quickly for the rest of us.

Arthur King, MD, US Army Surgeon, Medical Supply Officer-South Pacific. At my age, 94, my memory of some of those "recent" events is getting poor. I am now tied to oxygen ever since my bout with pneumonia and septicemia a few years ago. I won't be going anywhere near the South Pacific again soon.

William K. Schultz CM3/c, USN Malaria Control CMB538, SeaBees. I have not been back to the South Pacific since the war. The closest I have been is Hawaii, where I am going again next month. We usually go to the big island and will be spending a month there again. This time we will also stay a few days in Honolulu to see the battleship Missouri.

I think I would probably go back to some of the places I was in during the war. However, I'm sure things have changed by now so much that I would probably not recognize anything. The islands of the New Hebrides has some of the finest beaches in the world and really is quite pretty.

Thomas Gross, US Army 97th Field Artillery Battalion, Americal Division. I often thought of going back to the South Pacific with my wife. If money was no problem I would love to go back and see New Caledonia and Guadalcanal once more—to see where we lived and see if our bomb bunkers were still there and see what has changed.

Dave Minton, PhM1/c, USN Malaria Control Unit Cactus. I would never go back to the South Pacific. I spent the better part of two years down there and wouldn't consider going back for a visit even if someone gave me a free airline ticket. Korea, same thing, but that's another story. Bob and I had a job to do and we did it, simple as that. It was just a job like any other.

Lt. Cmdr. Lewis P. May, USNR Ret., Assistant Supply Officer, USS President Hayes APA-20. I am too old now too consider going back to the South Pacific, even though I still find the prospect of doing so very interesting.

Tom Croft, BM3/c USN, Gun Loader Turret-1, USS Monssen DD436. I often thought I'd like to see those islands again. I especially thought about it when those guys from National Geographic went down there and mapped the location of all those wrecks. I don't know about now, though, I'm really too old. You know I just turned 84 a couple of weeks ago.

Those islands were really beautiful, though. Every single one of them had a coconut grove on them. You know, I think it was Palmolive that owned the big one on Guadalcanal.

I would have liked to have seen some of those natives again, too. When I was down there and saw them, they were all wearing cut off shorts and really nothing else. They're probably a lot different today.

Marine Sergeant, Section Leader 75mm Pack Howitzers, I/3/11. Did you know that after the war, Lever Brothers, the company that owned the coconut plantation on the Canal, sued the U.S. Government for damages? Can you believe that? They sued the government because their coconut plantation was destroyed. They should have been glad just to get the place back. It's a strange world we live in, isn't it?

Robert Otto, BM3/c USN, 20mm Gunner USS Grayson. I never went back to the South Pacific. I thought about it many times, and still do. I often wonder how little Aore Island, in what was then the New Hebrides Islands, would look like now. There would be a world of difference I'm sure. Back then, most of the time I was on a ship with many shipmates. For eight months we saw the New Hebrides and Solomon Islands almost every day, but always from our ship. In late 1943 and early 1944, I left the Grayson for duty ashore at Aore. At one time, I knew that area very well. Again, you were always one small unit of a large group.

It is hard to explain, but many times I can picture myself leaning on the lifeline up near the bow of the ship as we moved along the coast of Guadalcanal on a quiet day. I am not so sure about sailing those waters again. Maybe I just do not want to pass over whatever remains of the Grayson's sister ships that might still be there. I will say this, at the time I was not sorry to leave the South Pacific. But had I known that I was headed for Iwo Jima and would go ashore with the Marines—well, you just go where your orders state.

Robert Michel, PhM1/c USN Veteran. You know, as I read over some of this stuff, I realize I have actually had quite an interesting life.

You want to know something? Every time I see any water, whether it's in a bucket or a barrel or a pond—wherever it is, do you know what I catch myself looking for, even today? That's right—larvae. And do you want to know something else? Most of the time they are there—floating right on top of the water, and I can still spot 'em.

Dennis Cline, author and student of the war in the South Pacific. It is unfortunate that comments from LtCmdr. James Sapero, Kenneth Knight, and Robert Mount have not been included in Bob's story—there is without a doubt so much that they could have contributed. All of us are aware of the effects time plays on our fragile existence. We should not forget, however, that it was these three men whose efforts throughout the war were instrumental in pioneering the efforts that ultimately saved thousands from the ravages of tropical disease. And, I might add, that Bob and Dave may well owe their very lives today to the attention paid them by their commanding officer, James Sapero, and their unit commander, Dr. Kenneth Knight.

Having spent the better part of a year reading every book I could get my hands on about World War II in the Pacific, and listening to the stories of so many of the various campaign's veterans, I can only say that, in some respects, I know less about the reality of the war against Japan than before I started. The stories told to me by Bob, Dave, John, Lew, Bill, William, Tom, Stewart, and all the others, painted a picture for me I had never before imagined.

I learned what it was like to sail on a troop ship, surviving for weeks on two meals a day, fighting for food so bad it was hard to look at, let alone eat. I tried to imagine what it would be like to be sent to a place I had never even heard of before. Though I had served in the Navy myself, at a much later date, I learned the true irony of ever volunteering for a job not clearly defined. I came to understand the deep emotions and absolute hatred these guys had for their enemy, and vice versa. I have a better understanding now why so many of our fathers and grandfathers have such a hard time with their children's acceptance of all things Japanese—cars, TVs, stereos, computers, and even our good old American work ethic. Hatred, bigotry, and cruelty are long remembered and often, too soon forgotten.

I am grateful to have been included, if just for a short time, in the lives of so many veterans and their families while gathering this material and retelling their stories. It has been an experience I will never forget. Only time and space have prevented me from presenting every one of their stories in the detail they really deserve. These men truly are heroes, and as such, their memories are deserving of our honor and respect. Thank you.

Chapter 15

Thoughts on the Literature of Guadalcanal and the South Pacific

When one considers the scope of the war in the South Pacific and the voluminous literature that has resulted, it is understandable how one might easily be overwhelmed. The reader is faced with hundreds of volumes, articles, and diaries covering a continuous series of campaigns that lasted from 1941 until 1945—a period of three years and which encompassed the efforts of virtually every branch of the military, on land, sea, and in the air.

Skeeter Beaters was never intended to be a historical treatise on the Solomons campaign, the siege of Guadalcanal, nor a detailed chronological description of events in the South Pacific. The story of the Skeeter Beaters and the resultant discussion of malaria control efforts was originally meant to be a round table conversation with a number of veterans who lived through very trying times and, when contacted to participate, were excited about sharing stories they still remember to this day. In fact, it seemed that the more time that was spent on the story, the more voluminous the outcome appeared. It is no wonder, then, that many a reader may be left asking more questions after reading this book than those who feel all of their questions have been answered. Consequently, it is this author's hope that some readers may have had their interest piqued enough to want to look deeper into the subject, either for more detail or a more diverse perspective. Thankfully, the subject of the South Pacific and the war in the Pacific is one that touched the lives of so many thousands that one need not look far to find an abundance of excellent books to choose from.

By conservative estimate, there are over one hundred books published, most still in print, telling the story of Guadalcanal and the Solomons campaign from every perspective one can imagine. All of the generals, admirals, and higher ranking military officers either published memoirs or biographies. The various branches of the military each released their history, which included in-depth accounts of their activities in the South Pacific, obviously from their unique perspective. Australian and British government employees who continued to operate on behalf of the Allies in the islands after the Japanese occupation of the area have provided us a number of excellent works on the campaign from the viewpoint of the Coast Watchers. Neither the air nor the naval contributions have been slighted in any way either, each category providing us with books that describe the men, their equipment, and the historic battles they fought. To gain an understanding of the Japanese perspective and detail of their participation will probably entail a little more work to find available material, but there are a few excellent books that have made it into English translation and are excellent additions to one's library.

The one book, however, necessary for anyone seeking a detailed chronology of the Guadalcanal campaign, and one that should be read prior to tackling any other on the subject, is Richard Frank's definitive work entitled **Guadalcanal** (Penguin Books, 1990). Frank's in-depth study ranks as the best over all historical reference available to date. His research and accuracy serve his readers well, telling the story of the war's most complex and hotly contested campaign that lasted well over six months. Frank's account includes all branches of the military, fairly presenting the perspcctive of both the U.S. and Japan. Early on in our narrative of the Skeeter Beaters, William Ricciardi reminded us that, "There was a lot of stuff going on down there that nobody ever tells you about." Frank does, as he walks his readers day by day through the give and take of combat on land, in the air and on the sea. Every unit, every ship, and every air action is meticulously detailed with supporting organization tables and attendant casualty reports.

For those just as interested in developing a good foundation for their exploration of the prelude to the war in the Pacific but don't fancy themselves as voracious readers, Dastar Corporation's Marathon Music and Video published an excellent, three volume work on video entitled "Guadalcanal: The Island of Death." This video is by far the best three hour program available. It begins with the Japanese attack on Pearl Harbor and continues through the pivotal naval battles of the Coral Sea and Midway. The three volume set provides the viewer an inside look at life on the island, from the embarkation of the troopships in August through the final evacuation of the Japanese in February 1943. The tapes treat the viewer to a skillful compilation of three historian's take on the land, sea, and air actions interspersed with actual footage and surviving veteran's commentary. Unlike other documentaries on the subject, "Guadalcanal: The Island of Death" not only gives a realistic taste of the hardships faced by the combatants, buts also offers some never before seen snapshots of the island's malaria control personnel in action, rare footage of the USS Monssen on coastal combat patrol, and equally scarce pictures of the Japanese enemy. "Guadalcanal: The Island of Death" is a production of unparalleled quality in content, realism, and style.

Now that we have developed a solid foundation with an understanding of the complexities of the events, chronology, and participants of the Solomon's campaign, we are prepared to tackle as much or as little of its nuance as the reader wishes by delving deeper into the available literature. The broad scope of the South Pacific campaign gives each of us the latitude to pursue the avenue most appealing to individual interests—air, naval or land actions, biographies, autobiographies, or even historic novels. Since **Skeeter Beaters** mostly covers the early participation of U.S. personnel in the South Pacific, our look at specific literature hereafter will reflect that perspective.

In a recent discussion with Dave Minton when we were kicking around the "loosey goosey" military set up in the South Pacific in the early years of the war, I made reference to a book by James Michener, **The Tales of the South Pacific.** This Pullitzer prize winning historic novel was published in 1946. I told Dave it was my opinion that it was that book alone which most closely approximated the discussions throughout **Skeeter Beaters**—more so than any other I had come across. Malaria control characters, earrings, the highly sought after pig tusks, the natives, the islands, "Bloody Mary" and the Tonkinese, and the boredom—a wide array of unbelievable stories that led me to believe that there was indeed a lot of crazy stuff going on down there at the time and that there were a lot of modern day pirates operating on behalf of the Allies.

Dave Minton, PhM3/c USN Malaria Control Unit Cactus I met Michener down there—I think it was at Espiritu. He was a Lieutenant Commander then. He showed me a one page article he had written. I think he said it was his first one. It was about a bunch of women—they were Polynesian or Melanesian, or some such, not the natives of the New Hebrides—that the French had hidden over on the Rennell Islands. They were special for some reason. I think that article was the Bali Hai he later wrote about.

The first book for one of America's most prolific authors, who went on to publish over eighty best sellers during his lifetime, "Tales of the South Pacific" deserves to be on every reader's list of South Pacific titles. Though first released at the end of the war, Michener's works continue on today in print and are mainstays on most neighborhood library shelves and in bookstores.

Two books, both written by participants of the barbaric ground actions on Guadalcanal, are "must-reads" for those who are half as inquisitive as I. **Helmet for My Pillow (Nelson, Doubleday, Inc., Bantam, 1979)** by Robert Leckie and **Guadalcanal Diary** (Random House Press, 1943) by Richard Tregaskis. Both narratives are autobiographical and provide the writers' personal perspective on the day to day strife faced on Guadalcanal by those who were there. Both books are today considered to be timeless classics of the period and each was subsequently adapted by Hollywood for the big screen.

Helmet for My Pillow is Leckie's own story of joining the United States Marine Corps at the outset of the war and his rites of passage at Parris Island Boot Camp, combat training at New River, and his subsequent involvement in the August 7th landing on Guadalcanal in the horror of the island's worst fighting. Another author whose first effort has earned its place on today's bookshelves, Leckie shares his life as a Marine, warts and all, taking us to Australia for the "great debauch" and finally to his last combat on Peleliu. Since the war, Robert Leckie has gone on to become a highly successful author, publishing more than a dozen books on American History. Over half of his work covers many of the most famous campaigns of his alma mater, the United States Marine Corps.

Guadalcanal Diary by Richard Tregaskis, on the other hand, is an interesting piece because it was written by a reporter who was sent to Guadalcanal to cover America's first offensive of the war. Told through the eyes of a trained observer, **Guadalcanal Diary**

became a best seller within days of its release in 1943. It still remains among the best accounts of Guadalcanal to this day, continuing to sell through numerous reprints. For the novice student of the campaign, however, **Guadalcanal Diary** may in some cases prove baffling. Time and places, as well as names and dates, are for the most part lacking, having been written during the war when such detail was considered vital to the security of our troops and was, therefore, subject to the censor's scissors. Though tough to follow without a basic knowledge of the campaign, Tragaskis' **Guadalcanal Diary** is an account that is so well told in such an easy style that it has found its way onto the reading list of many of our country's grade school history classes.

No well-rounded account of the war in the Solomons would be complete without looking at the lives of those most affected by its horror—the natives. It is easy for us to overlook the fact that over 100,000 native people were living on hundreds of islands in the area, large and small, when the Japanese first occupied them in the spring of 1942. What happened to these people, and how were their lives affected? When we read stories of Marines in hand-to-hand combat with the Japanese, or accounts of dogfights over the island, or the massive naval battles fought just offshore, it is easy to overlook the fact that native families long in residence still faced the day to day chore of raising their children, providing them food and shelter, and worrying about their well being, like parents anywhere else. Two little known books come to mind immediately to fill this void, both written by missionaries who each spent over twenty five years in the islands and who are well qualified to speak to us of the trials of their island brethren.

Half Pint on Guadalcanal by Sister Mary Paulita Schneller was published by her order, the Marist Missionary Sisters, in 1993. It is the biography of Sister Mary Evangeline, a young woman from Prince Edward Island who spent 53 years of her life in one of the world's most primitive environments. Arriving in the bay of Tulagi in 1929, Sister Mary Evangeline spent the next 13 years working to better the lives of her constituents across the channel on the island of Guadalcanal. When the Japanese arrived in 1942, Sister Evangeline was forced to flee to the interior of the island along with most of the native population. **Half Pint on Guadalcanal** is a rare inside look at civilian life during the deadly contest between the American liberators and the Japanese invaders. Sister Evangeline's story is not only one of courage and endurance, but one of the best chronicles of life "behind" the lines during the six months of the siege.

No story of Guadalcanal can be told without at least a cursory mention of the island's most famous and decorated son, Sir Jacob Vouza. Don Richter, a former Marine who saw action in the Pacific in World War II, went back to the islands after the war as a missionary, ministering to the archipelago's inhabitants for over 25 years. During his tenure, as he tells it, he became well acquainted with Jacob Vouza and his family and ultimately wrote the finest biography to date on his old friend. **Where the Sun Stood Still** (Toucan Press, 1992) is the definitive work on one of the campaigns greatest heroes and provides a necessary documentary of the participation of the Solomon Islanders in defense of their homeland against the Japanese invader. Again, it is easy for Americans to overlook the involvement of our Allies in the South Pacific, but it is books like this that remind us we were not alone when we faced the Japanese in the Pacific.

So much of the history of Guadalcanal and the South Pacific is clouded in the fog of legend and myth that one must not conclude any serious examination of the subject with reading only that literature which perpetuates the glory of the Marine Corps. Not to diminish the role and the sacrifice of the Marines, one is compelled to read of the Army's involvement as well. It was the Army and the 7th Marines who responded most rapidly to the threat of continued Japanese expansion in the South Pacific. The Army landed as early as April in New Caledonia; the 7th Marines took up defensive positions on Samoa. William McLaughlin was with one of those early arrivals with the Army, becoming one of the inaugural members of the soon to be formed Americal Division. In later years, McLaughlin wrote his autobiography entitled **The Americal Generation** (Cape Elizabeth Press). The book takes readers from New Caledonia through the final stages of the campaign on Guadalcanal, to reorganization on Fiji, to the bloody fighting on jungle-infested Bougainville, and concludes with the Philippines. McLaughlin's story gives us insight into a number of never before covered subjects such as his participation in combat with our Fijian allies on Bougainville to a day by day account of the bloody campaign for the Philippine Islands—both subjects rare to mainstream Pacific war literature. As a participant in three major campaigns, as well as a member of the Army of occupation in Japan, McLaughlin provides a necessary viewpoint very often overshadowed by the proliferation of perhaps more popular accounts of the war.

The deeper one delves into the subject of the war in the Pacific, the more one is struck with how horrifying the tale becomes. As the war progressed, U.S. forces advanced north and west up the South Pacific. They then crossed the Central and Northern Pacific for the final attack on the shores of the Japanese homeland itself. Many of our fighting men had by this time participated in as many as three, and often times even four, major landings. Or, they had fought in countless naval and air battles with literally no end in sight. The question for someone a generation removed from that time such as myself becomes, "How did any man continue on, battle after battle, with seemingly no end in sight? How could any man sustain any semblance of sanity without the least vestige of hope for survival?"

William Manchester, a combat Marine himself and in later years, the official biographer of Winston Churchill and Douglas MacArthur, went back to visit the Pacific battlefields in the late '70s to answer these and other questions that had plagued him over the years. The story of his travels to every major Pacific battleground is published in the book, **Goodbye, Darkness** (Doubleday, Bantam, Dell Publishing Group, 1979-80). This highly emotional narrative should be required reading for every student of the period. Step by step, Manchester visits every killing field on the road to Japan, recalling the bloody history at each stop and the emotional catharsis he experienced thirty-five years after the events had transpired. He visited places that are still today littered with the detritus of war, abandoned in the victor's haste to get back home. He stopped at cemeteries under the care of native islanders, and other areas once so remote, the average American couldn't even find them on a map. Yet today, these same areas are falling victim to the sprawl of suburbia that has taken over so much of the world. Islands, so far off the beaten track during the war, were now overrun with tourists. **Goodbye, Darkness** is the one book on the war that provides some sense of perspective and sanity to a time in our history not so long ago as to be forgotten, but far enough removed from our collective memory that it allows many of us to lose sight of the madness.

BIBLIOGRAPHY

Adam.com, Inc. My WebMD, Internet Source, 1999.

Altobello, Brian. Into The Shadows Furious. Novato, California: Presidio Press, 2000.

Armfield, M.A., Blanche B. Medical Department, United States Army in World War II: Organization and Administration in WWII. Washington, D.C.: Office of the Surgeon General, Department of the Army (Library of Congress Catalog Card Number 63-60002), 1963.

Asiapoint.net. Solomon Islands Profile. Insight Asia, 1998.

"Bora Bora," Microsoft Encarta Online Encyclopedia 2000, http://encarta.msn.com, 1997-2000.

"Bora Bora, The Marvelous Island." Tahiti Nui Travel, 2000.

Calhoun, C. Raymond. Tin Can Sailor: Life Aboard the USS Sterett, 1939-1945. Annapolis, MD: Naval Institute Press, 1993.

Churchill, Winston S. The Second World War, Volume IV, The Hinge of Fate. Boston, Mass.: Houghton-Mifflin Company, 1950.

City & County of San Francisco. History of the Treasure Island Project. 1999.

Croft, Tom. Author Interviews. July-November 2000.

Marathon Music & Video. Guadalcanal: The Island of Death. Dastar Corp., 1999.

Davis, Burke. Marine! The Life of Chesty Puller. Boston, Mass.: Little, Brown and Company, 1962.

Dickman, William C., Col. USMC, Retired. Author Interviews. October-December 2000.

Dower, John W. War Without Mercy: Race and Power in the Pacific War. New York: Pantheon Books, 1996.

Dyers, Vice Admiral George C. The Amphibians Came to Conquer. 1973.

Frank, Richard. Guadalcanal, The Definitive Account of the

Landmark Battle. New York: Penguin Books, 1990.
Frank, Richard. Downfall. New York: Random House, 1999.
Fuquea, Major David, C. Bougainville: The Amphibious Assault Enters Maturity. 1996.
Gross, Thomas R. Author Interviews. December 2000.
Gross, Thomas R. "I'll Go To Heaven Because I've Been In Hell." Unpublished diary. February 28, 1942-May 9, 1943.
Hacala, Mark T. USNR. Where It All Began: Hospital Corps Training Schools. 1999.
Halsey, Fleet Admiral William F. Memoirs.
Hammel, Eric. Guadalcanal: Starvation Island. New York: Crown Publishers, 1987.
Hiroyuki, Agawa. The Reluctant Admiral, Yamamoto and the Imperial Navy. Kodansha International, Ltd., 1979.
Hoffman, LtCol John T. USMCR. Chesty, The Story of Lieutenant General Lewis P. Puller, USMC. New York: Random House, 2001.
Horton, D.C. The Happy Isles: A Diary of the Solomons. London: Heinemann, 1965.
Horton, D. C. Fire over the Islands: The Coast Watchers of the Solomons. London: Leo Cooper, 1970.
Hutchinson Family Encyclopedia. Oxford, England: The Helicon Publishing, Ltd., 2000.
Imperial College of Science, Technology, & Medicine, Department of Chemistry. Drug Text.1993.
Johnson, Tim. Areca Catechu. Herbweb.com, 2000.
King, Arthur, MD. Author Interviews. January-February 2001.
King, Arthur, MD. Vignettes of the South Pacific, The Lighter Side of World War II. Bloomington, Indiana: Almagre Books, 1991.
Knight, Ruth. Author Interviews. April 2001 and September 2002.
Leckie, Robert. Helmet For My Pillow. New York: Nelson, Doubleday, Inc., Bantam Books, 1979.

Leckie, Robert. Strong Men Armed. New York: Da Capo Press, 1990.

Manchester, William. Goodbye, Darkness. New York: Bantam Doubleday Dell Publishing Group, Inc., 1979-80.

Marine Sergeant, 11th Marines. Author Interviews. November-December 2000.

May, Lewis P., Lt. Cmdr., USNR Retired. Author Interviews. December 2000.

McLaughlin, Bill. Author Interviews. January-February 2001.

McLaughlin, Bill. The Americal Generation. Cape Elizabeth Press, 1999.

Means, Lawrence. Author Interviews. October and November 2000.

Means, Lawrence. The Raggedy Ass Marines Are On Parade. Wildwood Publishing House, 1995.

Michel, Robert. Author Interviews. August 2000-June 2001.

Michener, James A. Tales of the South Pacific. 1974.

Miller, John M. Guadalcanal: The First Offensive, WWII. Barnes and Noble Books, 1995.

Minton, Dave. Author Interviews. August 2000-May 2001.

Ocean 98, Seahorse. "About Bora Bora." UN Seahorse, 2000.

Otto, Robert. Author Interviews. January-February 2001.

Resture, Jane. Jane's Oceania, 2000.

Ricciardi, William. Author Interviews. November-December 2000.

Richter, Don. Where the Sun Stood Still. Toucan Publishing, 1992.

Richter, John Francis. Author Interviews. November-December 2000.

Origins and Workings of the Aerosol Can. Robinson Research World of Knowledge. April 7, 1999.

San Francisco News. Treasure Island Goes Navy. April 17, 1942.

Schneller, Sister Mary Paulita. Half Pint On Guadalcanal.

Marist Missionary Sisters, 1993.
Shaw, Henry I. First Offensive: The Marine Campaign For Guadalcanal. Marines in World War II Commemorative Series, History and Museums Division, Headquarters U.S. Marine Corps., 1992.
Shaw, Henry I. Top of the Ladder: Marine Operations in the Northern Solomons. Marines in World War II Commemorative Series, History and Museums Division, Headquarters U.S. Marine Corps., 1997.
Stavisky, Samuel E. Marine Combat Correspondent: World War II In The Pacific. New York: Ballantine Publishing Group, 1999.
Steinert, David. The World War II Naval Corpsman, History of the Naval Hospital Corps. 2000.
Stinnett, Robert B. Day of Deceit: The Truth About FDR and Pearl Harbor. New York: Touchstone, Simon and Schuster, 2000.
Strickland, William K. Author Interviews. December 2000.
Taylor, Brian, PhD. Author Interviews. December 2000.
Tregaskis, Richard. Guadalcanal Diary. New York: Modern Library Paperback Press, 2000.
Twining, General Merrill B., USMC. No Bended Knee: The Battle for Guadalcanal, Memoir. Novado, California: Presidio Press, 1994.
Address of President Harry S. Truman. Bulletin 13, United States Department of State, September 2, 1945.
Navy Department Communiqués & Press Releases. U.S. Government Printing Office, 1943.
USS Monssen, Deck Logs: September 26-29, 1942. National Archives.
USS Monssen, War Diaries: September 9, 1942-October 7, 1942. National Archives, Office of Naval Records and Library.
van der Kolk, M.D., Bessel. "Posttraumatic Stress Disorder and

Memory." Psychiatric Times, March, 1997, Vol. XIV, Issue 3.
Westley, Colin. The Plain Man's Guide to Aerosols: The History of The Aerosol. YorksKaroo.net, 1998.
Winterton, Stewart. Author Interviews, June-September 2000.
WorldAtlas.com. "Bora Bora, Facts and Figures." Graphic Maps, 2000.

INDEX